Algebra

Algebra

Nicholas L. Pappas. Ph.D.

A Message about this Text: The subject is essentially endless. The purpose here is to say enough about the subject so that you, the reader, have a running start when you apply this knowledge to your work.

Prerequisites: Competence in Arithmetic

We believe important benefits accrue by doing the problems carefully, and by formulating the equations if only to copy them. These efforts in effect provide "startup" experience.

Once you have some experience we are confident that you will be able to expand your know how with reasonable effort.

A Message from the Author: I have worked continuously in the electronics industry since 1950 except for 11 semesters teaching at San Jose State University (Professor and Chair Computer Engineering 1988-1993). There I discovered my talent for teaching such as it may be. After War2 I attended Lehigh University, and then transferred to Stanford where I earned the MS degree and, while working at HP in the early 1950's, the Ph.D. EE degree. (Somehow I did not get the word and formally apply for the BS degree.) Hardware design has been my principal activity. I have learned enough about assembly language, Forth, C, and C++ to design the software I need for my projects.

Preface

This is about the fundamental ideas of Algebra, understanding why and how Algebra works.

The ideas of digit position and digit position weight are introduced to show how integers greater than 9 are created. In this way understanding[1] replaces rote learning. The real number system is reviewed.

Felix Klein[2] enumerates the eleven laws all elementary reckoning can be based on. There are five fundamental laws upon which addition depends. There are five exactly analogous laws upon which multiplication depends. And multiplication is connected to addition by the distributive law.

Fractions appeared when division created remainders. Fractions are numbers. Therefore fractions can be manipulated by the operations addition, multiplication, subtraction, and division.

Decimals are created when q divides p in the fraction p/q. The decimal point separates the integer part of a decimal from its fractional part. The ideas of digit position and digit position weight are extended past the decimal point to negative digit position numbers and fractional digit position weight. Decimals may be rational or irrational numbers.

A focus on general methods for solving algebraic equations allows one to know how to solve any problem. The numerous special methods have limited value.

Sometimes an equation is not in the desired form. Algebraic operations are used modify the form of the equation by making the *same* changes to both sides of =. The equality is not upset if the *same* changes are made to both sides of =.

A polynomial in one variable x is defined and its essential properties are discussed. The $+ \times - \div$ operations on polynomials are explained.

---

[1] N. L. Pappas *Arithmetic- Integers, Fractions, Decimals* ISBN 9781500104542
[2] Felix Klein 1908, "Arithmetic, Algebra, Analysis", ISBN 048643480X

Algebra

The Remainder Theorem is proved. How to find factors of polynomials and Newton's method for finding polynomial zeros are explained.

To solve an equation one finds all of its solutions. Cramer's Rule is the straightforward way to find solutions by determinants of algebraic equations. How to find solutions of linear equations by addition, subtraction or substitution is explained. The formula solving quadratic equations is derived and explained.

An *exponent n* is a symbol written above, and on the right of, another symbol known as the *base x* as in x^n. All arithmetic operations apply to exponents.

The Binomial Theorem shows how to expand $(a+x)^n$ when a and n are any numbers, positive, negative, integral or fractional. When a=1 expansions of $(1+x)^n$ provide useful approximations.

The Exponential b^x and Logarithmic Functions $\log_b x$ were created to solve problems not solvable by known functions. They are used in every branch of mathematics. From a numerical point of view logarithms may be the most important arithmetic concept in mathematics.

Many problems are simplified when a rational function, the ratio of two polynomials, is decomposed into a sum of partial fractions with denominators of lower degree. Partial fractions have many applications such as simplifying many algebraic problems as well as the inverse Laplace Transform process.

Matrix algebra allows one to write and process equations efficiently. Furthermore, in many problems, the matrix format makes the next step easier to perceive.

The mathematical induction method of *proof by induction* has many uses such as proving theorems, discovering new results, and providing relatively simple proofs of theorems obtained by other means.

A competent, serious, professional will tell you that acquiring knowledge is hard work. Specifically, many mathematicians will tell you that, even for them, learning and understanding mathematics at any level is not easy! Surprised?

Letters in Algebra In Arithmetic only numbers are used, because the goal is to learn how to manipulate numbers. Algebra is different. The goal is learning how to manipulate general equations.

The letters in the equations represent numbers. The letters represent words. We could write equations that are difficult-to-read using terms such as "first unknown", "second unknown", and so forth. Instead one letter words are used, words such as x, y, and z as in the equation $(x + y)(x - y) = x^2 - y^2$.

Please remember this When working with algebraic expressions, one is manipulating *numbers*. The rules for manipulating algebraic expressions are consistent with the properties of the *number system*. Therefore, when manipulating the symbolic quantities of algebra always ask yourself "If I replace the symbols by numbers in my results are the results valid?" Many times the answer is not obvious.

The text In this modest mathematics text, we have tried hard to write in plain English. We do not use the phrase *this is obvious* for a good reason. Nothing is obvious to a person learning any subject.

Most Algebra books are forests full of trees that make it very difficult for the reader to know what is important in a forest. This text contains very few trees.

Chapter abstracts are next.

> Our blog *npappasee.blogspot.com* may offer you additional information. Take a look.

> We would appreciate receiving your comments and views on this text at npappasz@yahoo.com.

Algebra

1 Integers The concepts of digit position and digit position weight are used to allow any base ten decimal number to be written using only the symbols 0 to 9. In fact any base may be used. Base 2 binary numbers are limited to symbols 0 and 1. The value of 10 depends on the number base.

2 Fractions Fractions appeared when division created remainders $(23/7=3+2/7)$. Fractions are numbers, which means fractions can be manipulated by the four operations addition, multiplication, subtraction, and division. The Greatest Common Divisor (gcd) finds the factors of two numbers. The Least Common Multiple (lcm) finds the smallest integer that can be divided by two integers. The gcd and lcm facilitate fraction operations.

3 Decimals Decimals are created when q divides p in the fraction p/q. The decimal point is a marker that separates the integer part of a decimal number from its fractional part. The ideas of digit position and digit position weight are extended to the right of the decimal point to negative digit position numbers and fractional digit position weights.

4 Algebraic Operations Algebraic operations use the four fundamental arithmetic operations plus the distributive law that connects addition to multiplication. Examples show how polynomial equations are modified to solve problems.

5 Polynomials The polynomial is the major form of algebraic expression. A polynomial in one variable x is defined. Presented are polynomial fundamentals such as operations on polynomials, the Remainder Theorem, factors of polynomials, the zeros of a polynomial, and Newton's method for finding polynomial zeros.

6 Polynomial Equations An *equality* is a statement that two algebraic expressions are equal. There are two kinds of equalities: identities and equations. We show that the roots of the polynomial f(x) in the equation f(x)=0 are solutions of that equation. Then we solve linear equations two ways: by elimination (addition and substitution), and by determinants using Cramer's Rule. Quadratic equations are solved by completing the square and by quadratic formula, which we derive. An example of a real electric circuit shows how algebra solves the circuit problem.

7 Exponents An *exponent n* is a symbol written above, and on the right of, another symbol known as the *base x* as in x^n. The expression x^n is referred to as a power; specifically the nth power of x. All arithmetic operations apply to exponents. *The base and exponent can be any type of numbers.*

8 The Binomial Theorem for any Index The Binomial Theorem shows how to expand $(x+a)^n$ when n is an integer thereby avoiding the tedious process of multiplying by (x+a) n−1 times. Furthermore, the Binomial Theorem shows how to expand $(x+a)^n$ when n is any number, positive, negative, integral or fractional. The Binomial Theorem has many applications such as calculating $(1+0.08)^4$ to many significant figures. Permutations and combinations facilitate the proof and applications of the Binomial Theorem.

9 The Exponential and Logarithmic Functions The Exponential b^x and Logarithmic $\log_b x$ functions were created to solve problems not solvable by known functions. They are used in every branch of mathematics.

Logarithms solve algebraic equations involving exponents. From a numerical point of view logarithms may be the most important arithmetic concept in mathematics.

10 Partial Fractions In many problems a rational function, the ratio of two polynomials, is decomposed into a sum of fractions with denominators of lower degree. Each fraction in the sum is referred to as a partial fraction. This process is a reverse of a process that adds fractions. Partial fractions have many applications such as simplifying many algebraic problems as well as the inverse Laplace Transform process.

11 Matrix Algebra The world says a matrix has rows and columns. A matrix is an array of r×c numbers, real or complex, arranged in r rows and c columns. Matrices allow one to write and process equations efficiently. Furthermore, in many problems, the matrix format makes the next step easier to perceive. This becomes clear as we proceed.

12 Mathematical Induction The Principal of Mathematical Induction is implemented as a mathematical formula involving the positive integer n that is true for all positive integers provided that (1) the formula is true when n=1, and (2) the hypothesis that the formula is true for any n is sufficient to ensure that the formula is true for n+1.

Contents

1 Integers

The world made a very wise decision when it decided not to introduce new symbols beyond the symbols 0 to 9. The world did not want to repeat the disaster of the Roman number system (such as XLVII or 47).

Consequently when the world tried to count units beyond 9 the world avoided introducing new symbols by introducing positions for digits to the left of the current digit. This means a number can have as many digits as we please so that we can count up to numbers such as 12, 237, 10009, and 123456789.

> As a result numbers based on the *ideas of position and position weight* only use the symbols 0 to 9.

Knowing how to count is a prerequisite to any mathematical activity. Counting is ground zero.

Any number represents some quantity of units. Know that the unit can be apples, cars, or whatever. The house is 27 feet wide. No, no the house is 9 yards wide. We choose one, one of anything, (whose symbol is 1) as our unit. Thus any number represents a quantity of ones. The symbol 5 represents the sum of five ones. Mathematics benefits immensely from abstractions such as 1, without committing to what 1 represents be it cars, centimeters, feet, or yards.

If you select any number, *the next number is found by adding 1.* In the abstract language of algebra, if letter n is shorthand for the word number and n represents any number, then the next number is $n+1$. This means there is no limit to the magnitude of a number.

Infinity is an alias for *a number that is as large as you please.*

To enable the discussion, we need to define *equals* (=) and *expression*. The = symbol connects two expressions such as A=B. The = symbol means that the expression A on the left hand side has the same value as the expression B on the right hand side.

Algebra

A mathematical expression is a group of symbols, numbers and/or operators, representing a quantity. In fact we can say that, in algebra, everything is an expression, and that not all expressions are equal.

There are four standard Arithmetic algorithms[1] for $+ - \times \div$ that convert one N-digit problem into N one-digit problems. If the algorithms were taught in schools, then understanding would replace rote learning. For example, here is an addition problem where N=5.

$$23641 + 18589 = (20000 + 10000) + (3000 + 8000) + (600 + 500) + (40 + 80) + (1 + 9)$$

The Real Number System

The real number system is a composite of the following subsets.

N The natural numbers
$\{1, 2, 3, 4, 5, \ldots\}$
This is the set of numbers used for counting (a.k.a counting numbers).

W The whole numbers
$\{0, 1, 2, 3, 4, 5, \ldots\}$
This set adds zero to the set of natural numbers.

Z The integers.
$\{\ldots, -6, -5, -4, -3, -2, -1, 0, 1, 2, 3, 4, 5, \ldots\}$
This set adds the negative integers to the set of whole numbers.

Q The rational numbers.
The set of rational numbers is the set of all numbers expressed as the ratio of two integers p/q where $q \neq 0$.
For example 1/3, 5/7, −213/5899, −2 (−2/1), etc

I The irrational numbers.
The set of all numbers whose decimal representations do not terminate nor repeat. This is the set of all numbers that is *not* expressed as the ratio of two integers p/q where $q \neq 0$.

---

[1] N. Pappas *Arithmetic- Integers, Fractions, Decimals* ISBN 9781500104542

1.1 Zero to Nine: 0 to 9, a review

Over the centuries people have agreed on names for numbers. There is, however, a very big problem with names, which are just words. The very big problem is that we cannot use words *to do* mathematics, because that would be tedious and probably impossible.

We can do mathematics if we use symbols in combinations. A study of the history of number reveals that the world has agreed on names and symbols for quantities ranging from zero to nine.

zero	*one*	*two*	*three*	*four*	*five*	*six*	*seven*	*eight*	*nine*
0	1	2	3	4	5	6	7	8	9

Number Definitions Definitions are implemented by adding 1 to produce the next number.

$0 = 0$

$1 = 1$

$2 = 1 + 1$

$3 = 1 + 1 + 1$

$4 = 1 + 1 + 1 + 1$

$5 = 1 + 1 + 1 + 1 + 1$

$6 = 1 + 1 + 1 + 1 + 1 + 1$

$7 = 1 + 1 + 1 + 1 + 1 + 1 + 1$

$8 = 1 + 1 + 1 + 1 + 1 + 1 + 1 + 1$

$9 = 1 + 1 + 1 + 1 + 1 + 1 + 1 + 1 + 1$

A *number line* is a geometrical representation of number. The very important number line is a graphic display of numbers.

The number line is constructed by marking off equal lengths along the line. Each mark on the number line is assigned a number. Assign 0 to any mark. Next, assign 1 to the first mark to the right of zero. Then the *distance* from 0 to 1 *represents* 1 unit of length. Subsequent marks to the right add 1 unit to the distance. Label subsequent marks 2, 3, 4, and up to 9. The question mark means we have run out of symbols.

1.2 Counting past 9, a review

When we count units, all we can do with units is to count from 0 to 9 over and over again. We *recycle* through 0, 1, 2, 3, ... , 9, 0, 1, 2, etc. Each time we reach 9 we have counted an additional 9+1 units.

This number 9+1 needs a name, and just how do we represent 9+1? Well, the number 9+1 has to be one *whatever*, and, if we continue counting. the next 9+1 makes it two whatevers. The whatevers have to be counted by another digit using 0 to 9, because no new symbols are allowed.

Another digit would allow us to add 1 to it each time the unit's count recycled from 9 back to 0 (another whatever). This other digit records the number of 9+1's. This digit needs a name. Someone must have said abracadabra and arbitrarily named the next number 9+1 *ten*, and so the other digit is the *ten's* digit. No new symbols means using only 0 to 9 as the ten's digit, All digits are 0 when we start counting.

```
0123456789  0123456789  0123456789  0123456789  012....  units
0000000000  1111111111  2222222222  3333333333  444....  tens
```

As we count, units increment from 0, and each time the units recycle from 9 back to 0 to start over, the ten's digit is incremented by 1 as shown here.

The situation, however, is unsatisfactory. Here is the unit's digit and over there is the ten's digit. What is a useful way to associate the unit and ten digits, so that we can calculate with numbers greater than 9?

After a long while, there was agreement to write numbers as if they were words. Agreement to write the digits in a sequence with the highest value digit first as in 10, which one could read as 1 ten and 0 units. In this way, ten became the *two* digit number 10 (one, zero).

| Now we know the question mark ? on the number line is a 10 |

We count from 0 to 9, 0 to 9, etc. Each time we count through 0, we increment the ten's digit as shown above. We continue counting until we reach 99 units, when we are faced with the question, what to do with 99+1?

We observe that 99+1 represents 9 tens + 9 units + 1 unit. We choose to convert the 9 units + 1 unit to 1 ten so that we can say 99+1 represents 9+1 or 10 tens. Consistent thinking produces yet another digit, a third digit to the left, that counts 9+1 tens. Since names are arbitrary we say abracadabra again and the name is *hundred's digit*.

This means when we count *tens* from 0 to 9, 0 to 9, etc., we increment the hundred's digit each time we count tens through 9 to 0. An important equivalent statement is we increment the hundred's digit each time we count tens/units up to 99, and recycle through 99 to 00. When tens and units make 99, and 1 is added, there is a roll over into hundreds.

0............9	0............9	0............9	0............9	0.......	*units*
0123456789	0123456789	0123456789	0123456789	012....	*tens*
0000000000	1111111111	2222222222	3333333333	444....	*hundreds*

The first time we count tens/units through 99 to 00, or 099 to 000, we increment the hundred's digit past 0. Consequently we write 99+1 in word format as 100 (one, zero, zero).

Can you guess what's next? We continue to count and soon we reach 999. The next number is 999+1. The name is thousand, and we write it in word format as 1000 (one, zero, zero, zero). When units, tens, and hundreds equal 9, then the next +1 causes a roll over into the next thousand.

0............9	0............9	0............9	0............9	0.......	*units*
0............9	0............9	0............9	0............9	0.......	*tens*
0123456789	0123456789	0123456789	0123456789	012....	*hundreds*
0000000000	1111111111	2222222222	3333333333	444....	*thousands*

In this way, counting units forces a counting of tens. And, in the same way, counting tens and units forces a counting of hundreds. And, again in the same way, counting hundreds and tens and units forces a counting of thousands. We are on our way to infinity.

1.3 Zero to Infinity

There is always a greater number. To any number n you add 1 to make the next number n+1.

For example, increase the group of 9999 by one unit. What do you get?

Nine thousand, nine hundred, ninety, nine, plus 1 is nine thousands, plus nine hundreds, plus nine tens, plus ten ones so you get ten thousands.

We write nine zero zero zero, 9 0 0 0 as 9000 for nine thousands. And then we write ten zero zero zero, 10 0 0 0, as 10000 for ten thousands. This is consistent with 10 0 0 or 1000, 10 0 or 100, and 1 0 or 10. Now we have a five digit number.

Add one to get ten thousand one, 10001.

Add one to 100 000 000 to get 100 000 001 (To be clear we have separated the digits into groups of three.)

Add one to 345 543 672 879 to get 345 543 672 880.

And on to infinity.

The genius of the decimal number system is that we only need the symbols 0 to 9, and the concept of position, to write any number we please, such as the numbers:

1876   314159   034   78241998543   3000000000001   4294967296

Then the concept of position weight gives each number a unique value.

1.4 Binary Numbers

We have been counting with ten different symbols. This system is the decimal number system with *base* ten. A number system can have *any* base you desire. As is well known, the base 2 (binary) number system is used by computers.

The one digit numbers used in the binary number system are zero (0) and one (1). Numbers from zero to infinity only use 0 and 1 such as 1010110. Confusion is avoided by adding subscript 2 as in 1010110_2.

When we try to count beyond 1 we run out of number symbols. We avoid new symbols when we replace the next number 1+1 by a digit that represents a quantity of 1+1 units. This new digit is named *two*. This is a $base_2$ system. The meaning of 10 (one, zero) is 2 in the $base_2$ system. So we have to know the base in order to know the value of 10 (one, zero).

The position 0 digit has weight 1 and the position 1 digit has weight 2. In the decimal system the position 0, 1, 2, 3, etc. weights are 1, 10, 100, 1000, etc. (multiples of 10). We infer that binary system positions 0, 1, 2, 3, etc. have weights that are 1, 2, 4, 8, etc. (multiples of 2).

Two makes sense, because we run out of symbols after two counts 0 to 1, whereas the decimal system runs out of symbols after ten counts 0 to 9. Two counts implies many digits in small binary numbers.

0	1	2	3	4	5	6	7	8	9	10	11	12	13	14	15	16
0	1	10	11	100	101	110	111	1000	1001	1010	1011	1100	1101	1101	1111	10000

Binary 1101 equals decimal 8+4+0+1=13.

The three columns of numbers show how binary numbers recycle. The first column recycles 0, 1. The second column recycles 00, 01, 10, 11 (0 to 3_{10}). The third column recycles 000 to 111 (0 to 7_{10}).

The rule is the same as the decimal rule. Increment the digit in position n+1 (the digit to the left) whenever all of the digits to the right of position n+1 rollover from 111… and recycle back to 0.

0	00	000
1	01	001
0	10	010
1	11	011
0	00	100
1	01	101
0	10	110
1	11	111

1.5 The Ideas of Position and Position Weight

The four digit number 1876 has four digits occupying four positions. Each position can only contain one digit, which is taken from the list of one digit numbers 0, 1, 2, 3, 4, 5, 6, 7, 8, 9. The four positions are filled by 1, 8, 7, and 6 respectively to create the number 1876.

We are not limited to four positions. We can have an unlimited number of positions. The rules are that any digit from 0 to 9 can be written in any position, and there can be as many positions as we desire. This is why we can write down any numbers we please such as these using only ten different symbols. We consider this to be a remarkable number system.

1876   314159   034   78241998543   3000000000001   4294967296

The positions are assigned numbers starting with 0 at the right hand digit[2]. The digits of 1876 are in positions 3, 2, 1, 0.

Now what? We need a new idea. In fact it was a *great* idea that someone unknown to us revealed a long time ago. The idea is assigning different weights for different positions in a number. (We used 1, 10, 100, ... weights when we counted past 9.)

When we count past 9 we have accumulated tens, hundreds, and so forth, which we report by incrementing the digit to the left when *all* digits to the right are 9. Each digit in a number is treated in the same way. This display shows hundreds incrementing by 1 as units and tens pass through 99 to 00.

0.............9   0.............9   0.............9   0.............9   0.......   *units* (*ten recycles*)
0123456789   0123456789   0123456789   0123456789   012....   *tens*
0000000000   1111111111   2222222222   3333333333   444....   *hundreds*

This is why giving each digit position a number and a weight is the key step in the process leading to larger numbers using only 0 to 9.

---

[2] The significance of defining position zero as zero instead of one becomes very clear when decimals are studied. There one learns about the idea of powers of ten and that 10_{10} to the zero power, 10^0, equals 1.

1.6 Position and Position weight revisited

Position weight implements the idea of having *different weights* such as 10000, 1000, 100, 10, 1 *for the positions of digits* in numbers. A necessary condition for the position weight idea to work is that any two adjacent columns have the same increase in weight. Observe that the weight of any two adjacent positions increases *times ten* when you move left one position.

Position #	8	....	4	3	2	1	0
Weight	100000000	....	10000	1000	100	10	1

Observe that the *number of zeros in the weight equals the position number*. Position 0's weight has no zeros, position 1's weight has 1 zero, and so forth.

When position weight increases from 1 to 10, 10 to 100, 100 to 1000, etc. as you move one column *to the left* of any column you are using a base 10 number system.

In position 0 the one digit numbers range from 0 to 9. The next number is 9+1, and is *defined* as the two digit number 10.

In position 1 the one digit numbers also range from 0 to 9. The next number is 9+1, and is *defined* as the three digit number 100.

In position 2 the one digit numbers also range from 0 to 9. The next number is 9+1, and is *defined* as the four digit number 1000.

And so on.

When you say two thousand ... thirty... four... as you look at the number 2034, you are using digit value. *By definition* the value of the 3 digit in position 1 is the sum of three position 1 weights of 10, which equals 30. The value of digits 2, 0, 3, 4 is 2000, 0, 30 and 4 according to position weight. This means we can expand 2034 as a sum of terms, which is 2000 + 000 + 30 + 4.

Emphasis: only one digit is allowed in each position.

Algebra

1.7 Composite and Prime Numbers

Multiplying numbers creates a product. The original numbers are *factors* of the product. Numbers w, x, y, z are factors of the product wxyz.

Numbers such as these are *composite* numbers.

$6 = 3 \times 2$
$12 = 2 \times 2 \times 3$
$150 = 5 \times 2 \times 5 \times 3$

Multiplication by 1 does not change a number. This is why 1 is referred to as a trivial factor.

Observe that there are numbers which do not have factors except the trivial factor 1. They are *prime* numbers.

$2, 3, 5, 7, 11, 13, 17, 19, 23, 29, 31, 37, \ldots$ *are primes*

You and the world are assured that *all composite numbers are products of primes*, because the Fundamental Theorem of Arithmetic has been proven to be true. The proof is presented when one studies Number Theory.

Fundamental Theorem of Arithmetic

> *For each integer $n > 1$ there exist primes*
> $p_1 \leq p_2 \leq p_3 \leq \cdots \leq p_r$ *such that*
> $n = p_1 \times p_2 \times p_3 \times \cdots \times p_r$
> *This factorization is unique*

1.8 The Basic Laws of Operations

Mathematics is about the reasoning, the ideas, not just the procedures.

Felix Klein[1] enumerates the eleven laws all elementary reckoning can be based on. There are *five fundamental laws* upon which *addition* depends. For any numbers x, y, z

(1) $x + y$ *is always a number* (*addition is always possible*)
(2) $x + y$ *is one valued*
(3)   *The associative law holds* $(x + y) + z = x + (y + z)$
(4)   *The commutative law holds* $x + y = y + x$
(5)   *The monotonic law holds*   *If* $y > z$, *then* $x + y > x + z$

There are *five exactly analogous laws* upon which *multiplication* depends.

(6) $x \times y$ *is always a number* (*multiplication is always possible*)
(7) $x \times y$ *is one valued*
(8)   *The associative law holds* $(x \times y) \times z = x \times (y \times z)$
(9)   *The commutative law holds* $x \times y = y \times x$
(10) *The monotonic law holds*   *If* $y > z$, *then* $x \times y > x \times z$

Multiplication is connected to addition by the distributive law.
(11) $x \times (y + z) = x \times y + x \times z$

Klein claims "that it is easy to show that all elementary reckoning can be based on these eleven laws." We take his word for it, because proving this statement requires a very big digression. Klein does refer to the content of original sources, which one can pursue. There are many reasons for knowing and understanding the eleven laws. Here are two.

1. We use them all of the time. Most of the time we use them without our being aware we are doing that, because their application is not explicit.
2. These laws play a central role in mathematics. They are hidden behind the scene, so to speak, of all operations as they are executed.

---

[1] Felix Klein 1908, "Arithmetic, Algebra, Analysis", ISBN 048643480X

2 Fractions

Any number is represented by a point on the number line. *Fraction* is the name of a number that may not be an integer. Nevertheless, fraction is a number that is on the number line. The *fraction*, a *rational number*, is written as *integer m over integer n*.

$\dfrac{m}{n}$ *is a simple fraction where m and n are any integers and* $n \neq 0$.

Important: in this chapter fraction components m and n are integers.

Fractions appeared when we studied division in Arithmetic.

A fraction can be located on the number line by using a straightforward geometrical construction that divides a line segment into n equal lengths. Integers mark off unit lengths on the number line. The geometrical construction shows how to divide any unit length into any number of equal lengths such as n=13, which shows there are 13 numbers in the length from marks 7 to 8 for example. In fact there is an infinity of numbers between any two integer marks (that's another subject). Fractions are some of that infinity of numbers, where we use infinity as an alias for *a number that is as large as we please*.

m may not be divisible by n A number m is not divisible by a number n when the result cannot be written as an integer. Not divisible means dividing m by n produces a quotient plus remainder, which produces a fraction such as 13/23.

$$\frac{dividend}{divisor} = quotient + \frac{remainder}{divisor} \qquad \frac{4521}{23} = 196 + \frac{13}{23} \quad and \quad 4521 = 196 \times 23 + 13$$

The remainder is zero when m is divisible by n. For example

$$\frac{dividend}{divisor} = quotient + \frac{remainder}{divisor} \qquad \frac{4508}{23} = 196 + \frac{0}{23} \quad and \quad 4508 = 196 \times 23 + 0$$

fraction operations Fractions are numbers. Therefore fractions can be manipulated by the operations addition, multiplication, subtraction, and division.

2.1 Fractions and the Number Line

Any number is a point on the number line. Integers are represented by marks on the number line, which are separated by equal unit lengths.

Divide each unit length into 7 equal lengths to produce a number line we label with two related scales.

The Standard Division Algorithm shows that 7 divided by 7 equals 1, and 14/7=2, 21/7=3. Now we can label the lower scale as shown, which we know is correct at integer marks 1, 2, 3. How about the other (fraction) marks?

The other marks are equidistant. Let the length equal x. Seven lengths x equal the length from 0 to 1, or 1 to 2, and so forth. The unit length equals the sum of seven 1/7 lengths.

$$7x = 1 \quad \Rightarrow \quad divide\ both\ sides\ by\ 7 \rightarrow \frac{7x}{7} = \frac{1}{7} \quad \Rightarrow \quad x = \frac{1}{7}$$

$$mx = m\frac{1}{7} = \frac{1}{7} + \frac{1}{7} + \frac{1}{7} + \cdots + \frac{1}{7} = \frac{1}{7}(1+1+1+\cdots+1) = m \times \frac{1}{7} \quad (m\ lengths\ \frac{1}{7})$$

The length 1/7 is the length from 0 to mark 1/7. Then 7 lengths 1/7 is the length from 0 to mark 7/7, which is the same as the length from 0 to 1. We can generalize from here and say *m lengths 1/n is the length from 0 to m/n.*

Names Names for a fraction's numbers are numerator and denominator.

$$fraction = \frac{numerator}{denominator} \quad such\ as \quad x = \frac{783}{147}$$

2.2 Comparing Fractions

Two fractions are two points on the number line. The fraction to the left on the line is less than the fraction to the right. Equal fractions are fractions that represent the same point on the number line. Equal fractions are also referred to as equivalent fractions. We prefer equal. When we multiply m/n by 1 the fraction's value does not change. Then if we replace the 1 by the fraction k/k and multiply, a fraction km/kn equal to m/n is created.

$$\frac{m}{n} = \frac{1 \times m}{1 \times n} = \frac{k \times m}{k \times n} = \frac{km}{kn} \quad \textit{for example} \quad \frac{7}{23} = \frac{1 \times 7}{1 \times 23} = \frac{6 \times 7}{6 \times 23} = \frac{42}{138}$$

Is 3/7 less than or greater than 5/11? The number line tells us its less.

Is fraction m/n less than (<), equal to (=), or greater than (>) fraction p/q? Comparing 3/7 and 5/11 is comparing apples and oranges. We can only compare apple-oranges to apple-oranges (fractions with the same denominators). Convert apples and oranges to apple-oranges by multiplying numerators and denominators as shown .

$$\frac{3}{7} = \frac{3}{7} \times \frac{11}{11} = \frac{33}{77} \quad \textit{and} \quad \frac{5}{11} = \frac{5}{11} \times \frac{7}{7} = \frac{35}{77} \quad \textit{and} \quad 33 < 35 \quad \rightarrow \quad \frac{3}{7} < \frac{5}{11}$$

Replacing numbers by variables, we produce a general solution. Convert apples (1/n) and oranges (1/q) to apple-oranges (1/nq) or (1/qn). Then compare mq to pn.

$$\frac{m}{n} = \frac{m}{n} \times \frac{q}{q} = \frac{mq}{nq} \quad \textit{and} \quad \frac{p}{q} = \frac{p}{q} \times \frac{n}{n} = \frac{pn}{qn}$$

The key to comparisons is to form equal denominators such as *nq*.

$$\textit{if the fractions are } \frac{m}{n} = \frac{3}{7} \textit{ and } \frac{p}{q} = \frac{5}{11} \textit{, then form } \frac{mq}{nq} = \frac{33}{77} \textit{ and } \frac{pn}{qn} = \frac{35}{77}$$

$$\textit{Now ask is } mq <, =, \textit{or} > pn? \qquad \textit{Is } 33 <, =, \textit{or} > 35?$$

2.3 Fraction Addition

Integers may be added if they have the same units of length. We cannot add 3 yards to 2 feet, but we can add 9 ft (3 yards) to 2 ft. The same is true of fractions. *One over the denominator n (1/n) is the fraction unit of length.* This is also true for integers where each integer n equals fraction n/1. Denominator 1 is the integer unit of length.

Fraction m/n is a quantity m of lengths 1/n (m/n = m × 1/n). Fraction 3/5 is a quantity 3 of lengths 1/5 (3/5 = 3 × 1/5). Fraction 11/17 is a quantity 11 of lengths 1/17 (11/17 = 11 × 1/17). We can add 3/5 (apples) to 11/17 (oranges) if we convert both to quantities of apple-oranges 1/(5×17) = 1/85.

> *The key to fraction calculations is "use the same units of length."*

Same denominators Fraction addition is straightforward when you want to add two fractions with the *same denominator d*, because each fraction is a quantity of the same length 1/d. In other words, calculations are facilitated, because the fraction 1/d with unit of length 1/d (an apple-orange) can be factored out. For example factor out 1/7:

$$\frac{4}{7}+\frac{2}{7}=4\times\frac{1}{7}+2\times\frac{1}{7}=(4+2)\times\frac{1}{7}=6\times\frac{1}{7}=\frac{6}{7}$$

If you interpret the fractions as 4 each 1/7 and 2 each 1/7, then you have a quantity of 6 each 1/7. Consequently you can use the shortcut: *if the denominators are equal, then add the numerators.* Verify on the number line by starting from 0 and laying out the lengths 4/7, and then 2/7.

We repeat. Observe that any fraction m/n = m × 1/n (e.g. 29/734 = 29 × 1/734). You can say the fraction m/n is a quantity m of the fraction 1/n. Thus you add quantities p and q of 1/n to get (p+q)/n. In general

$$\frac{p}{n}+\frac{q}{n}=p\frac{1}{n}+q\frac{1}{n}=(p+q)\frac{1}{n}=\frac{p+q}{n}\quad e.g.\quad \frac{29}{734}+\frac{214}{734}=\frac{29+214}{734}=\frac{243}{734}$$

Algebra

Different denominators Fraction addition is more complex when adding two fractions with *different denominators,* because this means adding quantities of different length such as 1/n and 1/q. The key step is converting both fractions to fractions that have the same unit of length 1/denominator.

$$\frac{m}{n}+\frac{p}{q}=\frac{m}{n}\times\frac{q}{q}+\frac{p}{q}\times\frac{n}{n}=\frac{mq}{nq}+\frac{pn}{qn}=mq\frac{1}{nq}+pn\frac{1}{qn}=(mq+pn)\frac{1}{nq}=\frac{mq+pn}{nq}$$

For example create two equal fractions each with a denominator equal to 12.

$$\frac{2}{3}+\frac{1}{4}=\frac{2}{3}\cdot\frac{4}{4}+\frac{1}{4}\cdot\frac{3}{3}=\frac{8}{12}+\frac{3}{12}=\frac{1}{12}(8+3)=\frac{11}{12}$$

Another example

$$\frac{7}{53}+\frac{22}{49}=\frac{7\cdot49}{53\cdot49}+\frac{22\cdot53}{53\cdot49}=\frac{343}{2597}+\frac{1166}{2597}=\frac{343+1166}{2597}=\frac{1509}{2597}$$

In other words, *cross multiply.* Multiply numerator 3 times denominator 7, and add the product of numerator 2 times denominator 8. Then multiply denominator 7 times denominator 8.

$$\frac{3}{8}+\frac{2}{7}=\frac{3\times7+2\times8}{7\times8}=\frac{21+16}{7\times8}=\frac{37}{56}$$

Emphasis: Study the following m and n details carefully to understand why we can cross multiply and avoid the details. The general case is

use $\dfrac{m}{m}=\dfrac{n}{n}=1$ *to add fractions with different denominators*

$$\frac{b}{n}+\frac{c}{m}=\frac{1}{mn}\times mn\left(\frac{b}{n}+\frac{c}{m}\right)=\frac{1}{mn}\left(\frac{mnb}{n}+\frac{mnc}{m}\right)=\frac{1}{mn}(mb+nc)=\frac{mb+nc}{mn}$$

Extra factors The following example brings up an important point. Extra factors are introduced unintentionally. How the *gcd* finds extra factors is explained in Section 2.7 page 21.

$$\frac{5}{6}+\frac{4}{9}=\frac{5}{6}\cdot\frac{9}{9}+\frac{4}{9}\cdot\frac{6}{6}=\frac{45+24}{54}=\frac{69}{54}$$

however $\dfrac{69}{54}=\dfrac{3\cdot23}{3\cdot18}=\dfrac{23}{18}$

24 Fraction Multiplication

Integer times integer Multiplication replaces sums with products.

$$m \times s = s + s + s + \cdots + s \quad (m \ terms)$$

Integer times fraction multiplication also replaces sums with products such as the sum of m lengths 1/n. Clearly multiplication by an integer means addition of many copies of a number such as a fraction.

$$m \times \frac{1}{n} = \frac{1}{n} + \frac{1}{n} + \frac{1}{n} + \cdots + \frac{1}{n} \quad (m \ terms) = \frac{1+1+1+\cdots+1}{n} = \frac{m}{n}$$

How is a fraction multiplied by a fraction implemented? Consider the area of the unit square with sides of length 1. The area of the unit square is 1×1=1. Divide the vertical y axis into 2 equal parts, and divide the horizontal x axis into 5 equal parts. This divides the unit square into ten equal areas whose total area equals 1. Therefore each of the ten equal areas have area 1/10, because a quantity 10 of 1/10 equals 10/10=1 the total area.

Each of the 10 areas has sides of length 1/2 and 1/5. The area of a rectangle is the product of its sides, and so each small area has area equal to 1/2 × 1/5. Therefore 1/2 × 1/5 must equal 1/10.

$$\frac{1}{2} \times \frac{1}{5} = \frac{1 \times 1}{2 \times 5} = \frac{1}{10}$$

What does 3/8 × 4/5 mean? For example the area of the 3/8 × 4/5 rectangle equals the area of 12 small areas where each small area has sides 1/8 and 1/5. Each small area equals 1/40, because

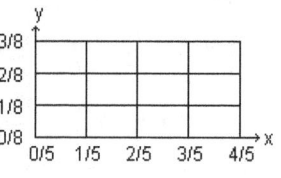

$$\frac{1}{8} \times \frac{1}{5} = \frac{5}{8 \cdot 5} \times \frac{8}{5 \cdot 8} = \frac{40}{40 \times 40} = \frac{1}{40} \implies area = 12 \times \frac{1}{40} = \frac{12}{40} = \frac{3 \times 4}{8 \times 5} = \frac{3}{8} \times \frac{4}{5}$$

17

Algebra

Emphasis: Multiplying a fraction by 1 does not change its value. And a 1=n/n where n is any number. To understand this cancel the n's to go back to 1. For example

$$\frac{1}{5} = \frac{1\times 7}{5\times 7} \quad \textit{because if the 7's are cancelled we are back to } \frac{1}{5}$$

In general

$$\frac{1}{n} = \frac{1\times p}{n\times p} = \frac{p}{np} \quad \textit{cancel the p's to return to } \frac{1}{n}$$

The examples show that a product of fractions is product of numerators over product of denominators.

$$\frac{m}{n}\times\frac{p}{q} = \frac{mp}{nq} = \frac{\textit{product of numerators}}{\textit{product of denominators}}$$

Here is how to multiply a fraction by an integer N.

$$N = \frac{N}{1} \quad\rightarrow\quad N\times\frac{p}{d} = \frac{N}{1}\times\frac{p}{d} = \frac{Np}{1d} = \frac{Np}{d}$$

A time saver cancels known common factors before multiplying. This saves a lot of work. For example:

$$\frac{12}{17}\times\frac{51}{28} = \frac{4\cdot 3}{17}\times\frac{3\cdot 17}{4\cdot 7} = \frac{3\cdot 3}{7} = \frac{9}{7}$$

A convention Fractions greater than 1 are referred to as *improper* fractions. There is a conventional writing of improper fractions greater than 1 that is not useful for calculations. The conventional writing is *not* a number based on position and weight (try adding two of them).

The convention is that you will see $2+\frac{2}{7}$ *written as* $2\frac{2}{7}$

To get a number do this $2\frac{2}{7} = 2+\frac{2}{7} = 2\times\frac{7}{7}+\frac{2}{7} = \frac{14}{7}+\frac{2}{7} = \frac{16}{7}$

We prefer to use the original fraction 16/7. We do not use the improper fraction convention, which leads to errors too easily.

2.5 Fraction Subtraction

Fraction subtraction is based on the idea of same units of length, which is used in fraction addition. A minus sign replaces the plus sign. In effect that is the only difference. For example:

$$\frac{4}{7}-\frac{2}{7}=\frac{1}{7}\cdot\frac{4}{1}-\frac{1}{7}\cdot\frac{2}{1}=\frac{1}{7}4-\frac{1}{7}2=\frac{1}{7}(4-2)=\frac{1}{7}2=\frac{2}{7}$$

Again, you can factor out the denominator every time if you create equivalent fractions. For example:

$$\frac{2}{3}-\frac{1}{4}=\frac{2}{3}\cdot\frac{4}{4}-\frac{1}{4}\cdot\frac{3}{3}=\frac{8}{12}-\frac{3}{12}=\frac{1}{12}(8-3)=\frac{5}{12}$$

Another way to get the answer saves some writing.

$$\frac{2}{3}-\frac{1}{4}=\frac{2}{3}\cdot\frac{4}{4}-\frac{1}{4}\cdot\frac{3}{3}=\frac{8-3}{12}=\frac{5}{12}$$

Another example brings up the same important point.

$$\frac{5}{6}-\frac{4}{9}=\frac{5}{6}\cdot\frac{9}{9}-\frac{4}{9}\cdot\frac{6}{6}=\frac{45-24}{54}=\frac{21}{54}$$

$$however \quad \frac{21}{54}=\frac{3\cdot7}{3\cdot18}=\frac{7}{18}$$

Sometimes one fraction is already in final form, and the result may be negative.

$$\frac{3}{16}-\frac{1}{4}=\frac{3}{16}-\frac{1}{4}\cdot\frac{4}{4}=\frac{3-4}{16}=-\frac{1}{16}$$

Algebra

2.6 Fraction Division

Fraction division calculations are totally different from integer division calculations. We show that the quotient of two fractions (m/n)/(p/q) is the fraction equal to the product of m/n and the *reciprocal* q/p of p/q. Formally

$$\frac{\dfrac{m}{n}}{\dfrac{p}{q}} = \frac{\dfrac{m}{n}\times\dfrac{q}{p}}{\dfrac{p}{q}\times\dfrac{q}{p}} = \frac{\dfrac{m}{n}\times\dfrac{q}{p}}{1} = \frac{m}{n}\times\frac{q}{p} = \frac{mq}{np}$$

So that once you have some experience.

$$\frac{\dfrac{num_1}{denom_1}}{\dfrac{num_2}{denom_2}} = \frac{\dfrac{11}{17}}{\dfrac{4}{3}} = \frac{11}{17}\cdot\frac{3}{4} = \frac{11}{17}\cdot\frac{3}{4} = \frac{33}{68} = \frac{num_1\cdot denom_2}{denom_1\cdot num_2}$$

A general argument for given two fractions x, y how do we calculate z=x/y?

$$let\ x = \frac{m}{n},\quad y = \frac{p}{q},\quad x = yz$$

$$the\ inverse\ of\ y\ is\ written\ as\ y^{-1}\ where\ y^{-1} = \frac{1}{y}$$

$$y^{-1}x = y^{-1}yz = z\quad because\quad y^{-1}y = \frac{1}{y}\times y = 1$$

$$z = xy^{-1} = \frac{m}{n}\times\frac{q}{p} = \frac{mq}{np}$$

Here are all of the steps, which you will actually skip once you have some experience.

$$\frac{\dfrac{a}{b}}{\dfrac{c}{d}} = \frac{\dfrac{a}{b}\cdot\dfrac{d}{c}}{\dfrac{c}{d}\cdot\dfrac{d}{c}} = \frac{\dfrac{a}{b}\cdot\dfrac{d}{c}}{1} = \frac{ad}{bc}$$

$$\frac{\dfrac{11}{17}}{\dfrac{4}{3}} = \frac{\dfrac{11}{17}\times\dfrac{3}{4}}{\dfrac{4}{3}\times\dfrac{3}{4}} = \frac{\dfrac{11}{17}\times\dfrac{3}{4}}{1} = \frac{11}{17}\times\frac{3}{4} = \frac{33}{68}$$

2.7 Greatest Common Divisor (gcd)

If you know the factors, a fraction can be simplified by *cancellation of the common factors* such as 3 and 5 in this example:

$$\frac{m}{n} = \frac{585}{165} = \frac{3\cdot3\cdot5\cdot13}{3\cdot5\cdot11} = \frac{3\cdot13}{11}\cdot\frac{3\cdot5}{3\cdot5} = \frac{3\cdot13}{11}\cdot\frac{3}{3}\cdot\frac{5}{5} = \frac{39}{11}$$

This can be difficult. For example what are the common factors of 69 and 1345, or of 4085376 and 297034? The gcd answers the question.

> The *greatest common divisor* (gcd) of two integers m and n is defined as the largest integer which divides both m and n (the gcd of m and n divides m and n).

Euclid's algorithm provides you with the means to find the gcd of any two integers m and n. The method is based on the fact that the last step of a sequence of division steps produces a zero remainder (that this always happens was proven by Euclid). You may remember how to find the gcd more readily if you note that the *next step is executing division of the reciprocal* of the remainder term of the present step. For example: 165/90 is the reciprocal of 90/165 (see below). We execute Euclid's algorithm on the number pair 585, 165. *The last divisor 15 is the greatest common divisor (gcd)*

$$\frac{m}{n} = q + \frac{r}{n} \quad \textit{where q is the quotient and r is the remainder}$$

$$\frac{m}{n} = q_1 + \frac{r_1}{n} \qquad \frac{585}{165} = 3 + \frac{90}{165} \qquad 585 = 165\cdot3 + 90$$

$$\frac{n}{r_1} = q_2 + \frac{r_2}{r_1} \qquad \frac{165}{90} = 1 + \frac{75}{90} \qquad 165 = 90\cdot1 + 75$$

$$\frac{r_1}{r_2} = q_3 + \frac{r_3}{r_2} \qquad \frac{90}{75} = 1 + \frac{15}{75} \qquad 90 = 75\cdot1 + 15$$

$$\frac{r_2}{r_3} = q_4 + \frac{r_4}{r_3} \qquad \frac{75}{15} = 5 + \frac{0}{15} \qquad 75 = 15\cdot5 + 0$$

The gcd of 585 and 165 is 15. Divide both numbers by the gcd to form the reduced fraction.

$$\frac{585}{15} = 39 \quad \textit{and} \quad \frac{165}{15} = 11 \quad \Rightarrow \quad \frac{585}{165} = \frac{15\times39}{15\times11} = \frac{39}{11}$$

2.8 Least Common Multiple (lcm)

The *least common multiple* of two integers m and n is the smallest integer which can be divided by both m and n (m and n divide the lcm).

$$m = 6, \ n = 9 \quad \Rightarrow \quad lcm = 18$$

because 18 is the smallest integer divisible by both 6 and 9

The lcm of two integers m and n is calculated as follows:

$$lcm = \frac{m \cdot n}{gcd \ of \ m \ and \ n}$$

if m = 6 and n = 9, then by the gcd algorithm

$$\frac{9}{6} = 1 + \frac{3}{6} \quad \Rightarrow \quad \frac{6}{3} = 2 + 0 \quad \textit{so that the gcd } = 3$$

and $lcm = \dfrac{m \cdot n}{gcd \ of \ m \ and \ n} = \dfrac{6 \cdot 9}{3} = \dfrac{3 \times 2 \cdot 3 \times 3}{3} = 3 \times 2 \times 3 = 18$

A practical method avoids calculating the gcd if you can factor m and n. In other words:

the factors of m and n are $6 = 3 \cdot 2$ and $9 = 3 \cdot 3$

the factors common to 6 and 9 are $2 \cdot 3 \cdot 3$

so that the lcm is $2 \cdot 3 \cdot 3 = 18$

The *important point* is that the step reducing the fraction is eliminated if the fractions have *denominators equal to the least common multiple*. Use n/n.

$$\frac{5}{6} + \frac{4}{9} = \frac{5}{6} \cdot \frac{3}{3} + \frac{4}{9} \cdot \frac{2}{2} = \frac{15 + 8}{18} = \frac{23}{18}$$

Sometimes one fraction is already in final form.

$$\frac{3}{16} + \frac{1}{4} = \frac{3}{16} + \frac{1}{4} \cdot \frac{4}{4} = \frac{3 + 4}{16} = \frac{7}{16}$$

Problems 2

The Basic Idea

An object n is divided into 23 parts.
1. What is one part called?
2. How is it written?

A number = 23/145.
3. Into how many parts is the original object divided?
4. How many parts of the original object does the fraction
  represent?

Equal Fractions

Find fractions that have a denominator of 128 equal to the following
fractions.
5. 1/4   6. 3/8   7. 7/16   8. 5/32   9. 17/64

Find fractions equal to the following pairs of fractions that have a
minimum common denominator. Hint - find factors of denominators.

10. $\dfrac{1}{4}\ \dfrac{3}{5}$ 11. $\dfrac{1}{14}\ \dfrac{3}{21}$ 12. $\dfrac{7}{15}\ \dfrac{5}{12}$ 13. $\dfrac{11}{16}\ \dfrac{3}{15}$ 14. $\dfrac{1}{7}\ \dfrac{3}{8}$

For each fraction find quotient plus remainder fraction.

15. 8/8	16. 9/8	17. 10/8	18. 11/8	19. 12/8	20. 13/8
21. 14/8	22. 15/8	23. 42/7	24. 22/7	25. 12/3	26. 14/3
27. 30/6	28. 33/6	29. 65/21	30. 27/11	31. 15/11	32. 19/4
33. 23/14	34. 45/9	35. 53/9	36. 27/5		

Change to larger denominators

37. 2/7 to 28ths	38. 3/5 to 20ths	39. 9/8 to 32ths	40. 3/10 to 1000ths
41. 11/6 to 72ths	42. 7/13 to 65ths	43. 2/9 to 27ths	44. /12 to 24ths

Algebra

Find the missing numbers.

45. $\dfrac{42}{7} = \dfrac{}{28}$

46. $\dfrac{2}{3} = \dfrac{}{27}$

47. $\dfrac{9}{4} = \dfrac{36}{}$

48. $\dfrac{}{6} = \dfrac{25}{30}$

49. $\dfrac{30}{} = \dfrac{5}{3}$

50. $\dfrac{33}{} = \dfrac{3}{7}$

51. $\dfrac{8}{21} = \dfrac{48}{}$

Fraction Addition

Add the following pairs of fractions. Hint - factor the denominators.

52. $\dfrac{8}{9} + \dfrac{1}{5}$

53. $\dfrac{3}{7} + \dfrac{1}{4}$

54. $\dfrac{5}{9} + \dfrac{3}{5}$

55. $\dfrac{9}{13} + \dfrac{2}{39}$

56. $\dfrac{2}{21} + \dfrac{3}{7}$

57. $\dfrac{4}{14} + \dfrac{1}{3}$

Convert to fractions and add.

58. $2\dfrac{7}{21} + 4\dfrac{5}{14}$

59. $3\dfrac{1}{12} + 5\dfrac{3}{18}$

60. $1\dfrac{3}{8} + 3\dfrac{5}{6}$

61. $6\dfrac{7}{13} + 1\dfrac{3}{26}$

62. $7\dfrac{7}{16} + 1\dfrac{11}{12}$

63. $4\dfrac{3}{11} + 1\dfrac{2}{9}$

Greatest Common Divisor (gcd)

Find the gcd of each pair of numbers.

64. 255 *and* 153

65. 336 *and* 280

66. 136 *and* 255

67. 105 *and* 168

Least Common Multiple (lcm)

Find the lcm of each pair of numbers.

68. 255 *and* 153

69. 336 *and* 280

70. 136 *and* 255

71. 105 *and* 168

Add the following pairs of fractions using the lcm

72. $\dfrac{13}{255} + \dfrac{7}{17}$

73. $\dfrac{4}{255} + \dfrac{5}{153}$

74. $\dfrac{9}{336} + \dfrac{2}{7}$

75. $\dfrac{13}{336} + \dfrac{3}{140}$

Find the lcm of

76. $\dfrac{1}{4}\ \dfrac{3}{5}\ \dfrac{7}{9}$

77. $\dfrac{2}{3}\ \dfrac{1}{15}\ \dfrac{9}{10}$

78. $\dfrac{5}{6}\ \dfrac{1}{2}\ \dfrac{1}{36}$

79. $\dfrac{3}{8}\ \dfrac{2}{9}\ \dfrac{1}{5}$

Fraction Subtraction

Subtract the fractions using lcm

80.　　　81.　　　82.　　　83.

$\dfrac{13}{85} - \dfrac{1}{17}$　　$\dfrac{37}{51} - \dfrac{5}{255}$　　$\dfrac{1}{14} - \dfrac{7}{168}$　　$\dfrac{13}{56} - \dfrac{3}{140}$

84. Decrease the value of 3/4 by twelve thirty ninths.
85. Decrease the value of 7/10 by two fifths.
86. Decrease the value of 4/9 by one fifth.
87. Decrease the value of 3/7 by one tenth.
88. Decrease the value of 17/32 by three eighths.
89. Decrease the value of 5/16 by two sevenths.

Fraction Multiplication

Multiply the fractions. Reduce to lowest terms. Hint - factor all numbers.

90.　　　91.　　　92.　　　93.

$\dfrac{3}{22} \cdot \dfrac{55}{17}$　　$\dfrac{18}{34} \cdot \dfrac{51}{28}$　　$\dfrac{84}{13} \cdot \dfrac{39}{132}$　　$\dfrac{28}{57} \cdot \dfrac{76}{7}$

94. Increase the value of 3/4 by four times.
95. Increase the value of 3/10 by two and one-half times.
96. Increase the value of 4/9 by five times.
97. Increase the value of 17/32 by two and two thirds times.
98. Increase the value of 4/9 by four and one half times.
99. Increase the value of 12/19 by one and one half times.

Find fraction to multiply by that decreases original fraction
100. Decrease 1/3 to 1/7.
101. Decrease 3/5 to 1/10.
102. Decrease 17/32 to 1/16.

Fraction Division

Divide the fractions. Reduce to lowest terms.

103　　　104　　　105　　　106

$\dfrac{\frac{3}{22}}{\frac{17}{55}}$　　$\dfrac{\frac{18}{34}}{\frac{28}{51}}$　　$\dfrac{\frac{13}{84}}{\frac{39}{132}}$　　$\dfrac{\frac{57}{28}}{\frac{76}{7}}$

Algebra

3 Decimals

Integers are whole numbers ranging in value from minus infinity to 0 to plus infinity, where infinity is an alias for a number as large as we please. Integers are represented by marks on the number line. Integer marks are separated by a distance equal to 1, as shown in this piece of the number line. The number line graphic shows that any number, such as x, is the sum of an integer (such as 4) and a part that is less than 1. The distance between the mark 4 and x is less than 1.

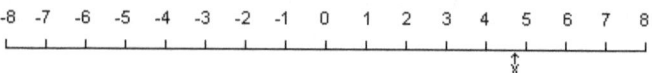

The integers mark off equal steps from minus infinity to positive infinity on the number line. The numbers in the gaps, such as x, between the integers 4 and 5 are *real* numbers, and so are the integers. The integers are a small subset of the real numbers. Some of the numbers in the gaps, such as 27+12/234, are fractions. Irrational numbers, such as √2, the square root of 2, fill in the remaining gaps. The numbers in the gaps have an integer part and a part that is less than 1. These numbers are written in different ways. We need a one way to write *any* number, which turns out to be the *decimal*. For example

$$27 + \frac{12}{234} = 27.05128.... \qquad\qquad \sqrt{2} = 1.41421....$$

A decimal representing any number has an integer part and a part that is less than 1. This is achieved by the decimal format *integer-dot-part<1* such as 627.23. The dot separates the parts. What does dot 23 (.23) mean?

The position weights of integer digits are 1, 10, 100, and so forth. The integer number 73841 has a digit in positions 4, 3, 2, 1, and 0. The corresponding position weights are 10000, 1000, 100, 10, 1. Position weight *decreases* by a factor of 10 as we move *right* one digit at a time.

Pretend a position exists to the right of units position 0. Then the weight of this position must also decrease by a factor of ten, because this is a move to the right.

Consequently the weight of this position has to be one tenth of 1 or the fraction 1/10. One more move to the right has to be 1/10 of 1/10 or 1/100 (1/100 = 1/10×1/10).

For example, the fraction 23/100=20/100+3/100. This fraction reduces to 2/10+3/100, which we rewrite as (2×1/10)+(3×1/100) in anticipation of writing a decimal number.

The decimal number representing the fraction 23/100 is written as 0.23. The *dot* before the fractional part 23 differentiates the fraction 0.23 from the integer 23. This *dot* notation is referred to as the *decimal point*.

The mathematical community has agreed to write fraction 23/100 as 0.23. The decimal number 0.23 means zero integer part plus a 23/100 fraction part. Observe that the part less than 1, .23, is written as if it is an integer using only the ten symbols 0 to 9. However the dot before the 23 changes the meaning from 23 to (2×1/10)+(3×1/100)=23/100.

Decimal point The decimal point in 0.23, marks the boundary between integer 0 and fraction 23/100. *A decimal point is simply a marker.*

Position number extended The digits in the number 43210.12345 are equal to the *magnitude* of their position number. Positions to the right of the decimal point have negative numbers that start from −1. Why this is so is explained in upcoming paragraphs. Emphasis: *the decimal point is simply a marker.*

Numbers such as 65.23 are called decimals.

Decimals are numbers you can add, subtract, multiply, and divide. There are many applications of decimals in science and commerce.

Read the decimal fraction 12.43 as "twelve and forty three one hundreds" (1/100). Or read it as "12 point 43."

> *The essential idea is the dot marking the point separating the integer part that is greater than or equal to 0, from the part that is 0 or less than 1 as in 5.321, 5.0, 0.0, 0.321*

3.1 Powers of Ten and the Base 10 Number system

An *exponent x* is a symbol written above, and on the right of, another symbol known as the *base b* as in b^x (Chapter 9). The expression b^x is referred to as a power; specifically the xth power of b. All arithmetic operations apply to exponents. *The exponent can be any type of number, which we restrict to integers* in this chapter. Furthermore let b=10 so that we can discuss powers of ten such as 10^5.

Many famous physical constants are usually expressed as a number in the range 1 to 9.999..... times a power of 10. Observe the negative exponents.

velocity of light $c = 2.997925 \times 10^8 \, meters/second$

Avogadro's number $N_A = 6.0225 \times 10^{23} mole^{-1}$

charge of the electron $e = 1.60210 \times 10^{-19} Coulomb$

Planck's constant $h = 6.62517 \times 10^{-34} Joule \cdot second$

Boltzman's constant $k = 1.3805 \times 10^{-23} Joule/degree$

Position weights are 1, 10, 100, 1000, etc., which are products of tens. This means the weights can be represented by a power of ten such as 10^p where the position number p is the exponent, and the base is 10.

We are interested in powers of ten, because ten is the base of our decimal number system. We have expanded integers as the sum of the value of their digits. For example the whole number $596 = 500 + 90 + 6$

We know we can do this, because when we learned to count we learned that for 596

- Digit 6 is in position zero, and that a digit equal to 6 in position 0 represents a quantity of six 1's.
- Digit 9 is in position one, and that a digit equal to 9 in position 1 represents a quantity of nine 10's.
- Digit 5 is in position two, and that a digit equal to 5 in position 2 represents a quantity of five 100's.

The idea of position increases the value of a digit *to the left* by a factor of ten. This is why the weight of any position is a power of ten.

Consider a conceptual explanation of a zero or negative exponent.

$exponents\ add:$ $\quad 10^{2+3} = (10 \cdot 10) \times (10 \cdot 10 \cdot 10) = 10^2 \times 10^3 = 10^5$

$exponents\ subtract:$ $\dfrac{10^5}{10^2} = \dfrac{10 \cdot 10 \cdot 10 \cdot 10 \cdot 10}{10 \cdot 10} = \dfrac{10^5}{10^2} \times \dfrac{10^{-2}}{10^{-2}} = 10^{5-2} = 10^3$

$specific\ case:$ $\dfrac{10^0}{10^1} = 10^{0-1} = 10^{-1} = \dfrac{1}{10}$ $\quad \Rightarrow \quad \dfrac{10^0}{10^1} = \dfrac{1}{10}$ $\quad \Rightarrow \quad 10^0 = 1$

Since 10 with exponent 0 equals 1, $10^0=1$, the *position number* of the units digit was given the number zero.

This is why exponents of position weights equal position number. For integers:

position#	position wt	exponent
0	$1 = 10^0$	0
1	$10 = 10^1$	1
2	$100 = 10^2$	2
3	$1000 = 10^3$	3

Note : position number increases by one for each move to the left.

Pretend you are in position 3. Position number decreases by one as you move *to the right* from position to position 3, 2, 1, 0. You reach the units digit in position 0. Then the digit *to the right* of the units digit must be in position 0–1 or −1 (minus one). For fractions we have

position#	position weight	exponent
−1	$\dfrac{1}{10} = \dfrac{1}{10^1} = 10^{-1}$	−1
−2	$\dfrac{1}{100} = \dfrac{1}{10^2} = 10^{-2}$	−2
−3	$\dfrac{1}{1000} = \dfrac{1}{10^3} = 10^{-3}$	−3

Now merge whole number positions with fraction number positions.

position#	3	2	1	0	−1	−2	−3
weight	1000	100	10	1	$\dfrac{1}{10}$	$\dfrac{1}{100}$	$\dfrac{1}{1000}$
		10^3	10^2	10^1	10^0	10^{-1}	10^{-2}	10^{-3}
exponent	3	2	1	0	−1	−2	−3

Algebra

3.2 Decimal point and position weight

The logic of powers of ten places the integer part to the left of the decimal point, and the fraction part to the right of the decimal point. The decimal point separates integer digits from fractional digits. Integer digits have values based on positive powers of ten, whereas fractional digits have values based on negative powers of ten. For example:

$63.71 = 6 \cdot 10^1 + 3 \cdot 10^0 + 7 \cdot 10^{-1} + 1 \cdot 10^{-2}$

A more complex example is the number 210.1234. In this made up number the digits correspond to the position numbers, which in fact are actually −1, −2, −3, −4 to the right of the decimal point. The position number *sign* is positive for digits to the left of the decimal point, and negative for digits to the right of the decimal point. We expand 210.1234 as a sum of terms, and convert each term to a powers of ten format.

$210.1234 = 200.0 + 10.0 + 0.0 + 0.1 + 0.02 + 0.003 + 0.0004$

Check by adding the terms. Align the decimal points so that digits

with same weight are added

200.0

  10.0

    0.0

    0.1

    0.02

    0.003

    0.0004

Convert terms of 210.1234 to powers of ten.

$210.1234 = 2 \cdot 100 + 1 \cdot 10 + 0 \cdot 1 + 1 \cdot \dfrac{1}{10} + 2 \cdot \dfrac{1}{100} + 3 \cdot \dfrac{1}{1000} + 4 \cdot \dfrac{1}{10000}$

$210.1234 = 2 \cdot 100 + 1 \cdot 10 + 0 \cdot 1 + 1 \cdot \dfrac{1}{10^1} + 2 \cdot \dfrac{1}{10^2} + 3 \cdot \dfrac{1}{10^3} + 4 \cdot \dfrac{1}{10^4}$

$210.1234 = 2 \cdot 10^2 + 1 \cdot 10^1 + 0 \cdot 10^0 + 1 \cdot 10^{-1} + 2 \cdot 10^{-2} + 3 \cdot 10^{-3} + 4 \cdot 10^{-4}$

The unwritten decimal point falls between the 0 and −1 power terms (positions 0 and −1). The *decimal point is not present* in the expansion as a sum of powers of ten, because it only marks the boundary between the whole and fraction parts when the number is written as a decimal.

3.3 Conversions to Decimals

The decimal forms of numbers are essentially mandatory for computational purposes. A decimal is a number that includes a decimal point. A decimal is usually the sum of an integer and a fraction: e.g. 46.9835. The fraction or integer may be zero as in 25 = 25.0, or 1/2 = 0.5. A decimal point is the dot marking the boundary separating the integer part from the fraction part.

Convert an Integer Any integer can be converted to decimal form by adding zero value in the form of a dot and zero(s). For example 43 = 43.0 = 43.000, etc.

Convert a fraction to a terminating decimal A way to convert some, *but not all*, fractions into a decimal is to make the denominator a power of ten. We multiply the fraction by n/n where the integer n times the denominator is some power of ten. You may want to refer to Section 3.6 to understand the following two examples, because here and there we *move* the decimal point as we change powers of ten.

Create an equivalent fraction with power of ten denominator

$$\frac{1}{25} = \frac{1}{25} \times \frac{4}{4} = \frac{4}{100} = 4 \cdot 10^{-2} = 0.4 \cdot 10^{-1} = 0.04 \cdot 10^{0} = 0.04$$

Convert 8 to a power of ten. Multiply by $1 = \frac{n}{n} = \frac{125}{125}$

$$\frac{7}{8} = \frac{7}{8} \times \frac{125}{125} = \frac{875}{1000} = 875 \cdot 10^{-3} = 0.875$$

Convert a fraction by division Any fraction is converted to a decimal by the division. The new item is keeping track of the decimal point, which is a straightforward procedure.

Convert fraction 5/8 to 0.6250 The remainder is zero after the fourth division, which means the decimal terminates or repeats the 0 in this example. The ellipses ... mean that the number 0, in this example, repeats forever.

Convert fraction 2/11 to 0.1818 The remainders alternate: 2, 9, 2, 9, etc. The non zero remainders means the division can continue forever. The quotient digits 1, 8 repeat forever.

Algebra

Convert fraction 1/7 to 0.142857 The remainders recycle through 1, 3, 2, 6, 4, 5. The non zero remainders means the division continues forever. The number 142857 repeats forever.

```
            0.142857...
         7)1.000000...
          -0
           1 0
          - 7
            30
          - 28
            20
          - 14
            60
          - 56
            40
          - 35
            50
          - 49
             1...
```

```
    0.62500...
  8)5.00000...
   -0
    5 0
  -4 8
    20
  -16
    40
  -40
    00
  -00
     0...
```

```
    0.1818...
 11)2.0000...
   -0
    2 0
  -11
    90
  -88
    20
    11
    90
  -88
    2...
```

0.6250 is a terminating decimal
0.181818... is a repeating decimal repeating 18.
0.142857142857... is a repeating decimal repeating 142857.

Note: *The decimal form of irrational numbers does not terminate, nor repeat.* Here are 9 digit approximations of three irrational numbers.

$\pi \approx 3.14159265$ $\sqrt{2} \approx 1.41421356$ $\sqrt{3} \approx 1.73205081$

Theory of the Repeating Decimal

Felix Klein[1] offers this argument.

Theorem of Fermat   for every prime number p, except 2 and 5,

$$10^{p-1} \equiv 1 (\text{mod } p)$$

this means $\dfrac{10^{p-1}-1}{p} = \dfrac{10^{p-1}}{p} - \dfrac{1}{p} = N$ *integer*

Another theorem from the theory of numbers is the theorem that the smallest exponent δ is either p–1 or a divisor of p–1.

$$\dfrac{10^{\delta}-1}{p} = N \implies \dfrac{10^{\delta}}{p} = N + \dfrac{1}{p} \implies \dfrac{10^{\delta}}{p} \text{ and } \dfrac{1}{p} \text{ differ by an integer } N$$

this means the fractional part of $\dfrac{10^{\delta}}{p}$ *equals* $\dfrac{1}{p}$

This is the key idea: *Decimal numbers that differ by an integer have equal decimal fraction parts.*

Think of 1/p as a decimal. Then $10^{\delta} \times$ 1/p only moves the decimal point δ positions to the right. Consequently the digits are not changed.

We assume the decimal fraction has a repeated *period*, or block, of δ digits.

Examples showing δ may be less than p as well as equal to p–1 (Section 3.4).

$\dfrac{1}{3} = 0.33333....$ (δ = 1)

$\dfrac{1}{11} = 0.090909....$ (δ = 2)

$\dfrac{1}{7} = 0.142857142857....$ (δ = 6)

---

[1] Felix Klein 1908, "Arithmetic, Algebra, Analysis" , ISBN 048643480X

3.4 Decimals to Fractions

The Standard Division Algorithm[2] was used to show that any rational number, such as a fraction m/n, has one block of digits or a repeating block of digits in its decimal fraction.

Now we want to show that any decimal with a repeating block equals a rational number. The key idea stems from the fact that decimal numbers that differ by an integer have equal decimal fraction parts (such as 23.987 and 654.987).

Given a decimal fraction with a repeating block of δ digits multiply it by 10^δ to leave the fraction part unchanged thereby creating a decimal number that differs from the decimal fraction by an integer. Then subtract the original decimal fraction to get an integer (Theory of the Repeating Decimal page 33). Three examples.

$let\ x = 0.33333....\quad (\delta = 1)$

$10^1 x = 3.33333....\ integer\ 3\ plus\ a\ fraction\ part\ that\ is\ not\ changed$

$10^1 x - x = 3.33333....- 0.33333....$

$9x = 3.33333....- 0.33333....$

$9x = 3 \qquad \Rightarrow \qquad x = \dfrac{1}{3}$

$let\ x = 0.090909....\quad (\delta = 2)$

$10^2 x = 09.090909....$

$10^2 x - x = 9.090909....- 0.090909....$

$99x = 9.090909....- 0.090909....$

$99x = 9 \qquad \Rightarrow \qquad x = \dfrac{1}{11}$

$let\ x = 0.142857....\quad (\delta = 6)$

$10^6 x = 142857.142857....$

$10^6 x - x = 142857.142857....- 0.142857....$

$999999x = 142857.142857....- 0.142857....$

$999999x = 142857 \qquad \Rightarrow \qquad x = \dfrac{1}{7}$

---

[2] N. Pappas *Arithmetic- Integers, Fractions, Decimals* ISBN 9781500104542

3.5 Decimal Operations

Decimals are numbers that can be operated on by the addition, subtraction, multiplication and division operators. When we added or subtracted integers we first aligned their right-hand edges.

4567 *right - hand edges aligned*

+234
_____

4801

In fact we were aligning integers at their unwritten decimal points when we added or subtracted.

4567.0 *decimal points aligned*

+234.0
_____

4801.0

The alignment by right-hand edge or (unwritten) decimal point is alignment by position, which is necessary in order to add digits with the same position weight. The alignment by decimal points is equivalent to alignment by position. This is why the introduction of decimal points does not affect calculations using integers. The introduction of decimal points extends calculations with integers to calculations with any real numbers.

Add and subtract After aligning decimal points, ignore the decimal points when adding or subtracting.

234.567	234.567
+98.7	−98.700
333.267	135.867

Algebra

Multiply Pretend the decimal points are not there, and just multiply the two numbers. Then insert a decimal point in the answer s positions to the left. You determine s by counting digits to the right of the decimal points in both numbers. We show that this is correct by example.

Multiply 435.56 by 34.567 while ignoring the decimal points to get 1,505,600,252. There are 2 + 3 or 5 digits to the right of the decimal points, which makes s = 5. The answer is 15056.00252. Verify by rounding down and up to get an estimate of 435 × 35 = 15225, which is the same order of magnitude as 15056.

$$
\begin{array}{r}
43556 \\
\times 34567 \\
\hline
304892 \\
2613360 \\
21778000 \\
174224000 \\
+1306680000 \\
\hline
1505600252
\end{array}
$$

$$435.56 \times 34.567 = 15056.00252$$

Divide Convert the divisor to an integer and use conventional division. In the case of 234.567/98.7 convert 98.7 to 987. In other words multiply both numbers by ten, This moves the decimal point one place to the right in both numbers (see 3.6). Then divide, while pretending the decimal point is not there.

$$
\begin{array}{r}
2.37\ldots\ldots \\
987\overline{)2345.670000} \\
\underline{1974} \\
3716 \\
\underline{2961} \\
7557 \\
\underline{6909} \\
6480 \ etc
\end{array}
$$

3.6 Moving the decimal point left and right

***A move to the right means times* 10** If we rewrite the number 210.123 by moving the decimal point one place to the right, then the number changes to 2101.23. The rewritten value is ten times the original value.

$$2101.23 = 2\cdot10^3 + 1\cdot10^2 + 0\cdot10^1 + 1\cdot10^0 + 2\cdot10^{-1} + 3\cdot10^{-2}$$
$$= 10\cdot(2\cdot10^2 + 1\cdot10^1 + 0\cdot10^0 + 1\cdot10^{-1} + 2\cdot10^{-2} + 3\cdot10^{-3})$$
$$= 10 \times 210.123$$

> *Moving the decimal point in any number, such as 210.123, one position to the right is the same as multiplying by ten.*

For example:

Every multiplication by ten moves the decimal point to the right
$$5.0\times1000 = 50.0\times100 = 500.0\times10 = 5000.0\times1 = 5000$$

***A move to the left means times* 1/10** If we rewrite the number 210.123 by moving the decimal point one place to the left then the number changes to 21.0123. The rewritten value is one–tenth the original value.

$$21.0123 = 2\cdot10^1 + 1\cdot10^0 + 0\cdot10^{-1} + 1\cdot10^{-2} + 2\cdot10^{-3} + 3\cdot10^{-4}$$
$$= \frac{1}{10}(2\cdot10^2 + 1\cdot10^1 + 0\cdot10^0 + 1\cdot10^{-1} + 2\cdot10^{-2} + 3\cdot10^{-3})$$
$$= \frac{1}{10} \times 210.123$$

> *Moving the decimal point one position to the left in any number, such as 210.123, is the same as dividing by ten.*

For example:

Every division by ten moves the decimal point to the left
$$\frac{500}{1000} = 500.0\times\frac{1}{1000} = 50.0\times\frac{1}{100} = 5.0\times\frac{1}{10} = 0.5$$

Problems 3

Definitions
1. The denominator of a decimal fraction is a power of what number?
2. What is the purpose of the decimal point?
3. A mixed number is the sum of two parts. What are their names?
4. The decimal point separates two parts of any number. What are their names?

Conversions to decimal form Convert to a decimal.

5.	6.	7.	8.	9.	10.	11.	12.	13.	14.	15.
$\dfrac{3}{6}$	$\dfrac{3}{7}$	$\dfrac{5}{20}$	$\dfrac{13}{17}$	$\dfrac{17}{123}$	$\dfrac{5}{16}$	$\dfrac{1}{6}$	$\dfrac{1}{11}$	$\dfrac{13}{45}$	$\dfrac{23}{54}$	$\dfrac{45}{111}$

Convert the following phrases into decimals.
16. One-hundred-three one-thousandths
17. Thirty-nine one-thousandths
18. Nine one-hundredths

Decimal point and position weight
Expand the numbers into a sum of powers of ten. Omit the zero terms.

19.	20.	21.	22.
5.3	347.59	502.01	20001.0001

Moving decimal point left and right
23. What operation with what number moves a decimal point to the left?
24. What operation with what number moves a decimal point to the right?

Leading and trailing zeros
25. Adding a leading zero to the integer part with value x changes the value to what?
26. Adding a trailing zero to the integer part with value x changes the value to what?
27. Adding a leading zero to the fractional part with value x changes the value to what?
28. Adding a trailing zero to the fractional part with value x changes the value to what?

Tenths, hundreds, thousandths

29. Write thirty seven hundredths in numerical format.
30. Write thirteen tenths in numerical format.
31. Write fifty seven thousandths in numerical format.
32. Write one hundred ninety nine ten thousandths in numerical format.
33. Write one one millionth in numerical format.

Rational numbers and repeating decimals

34. Find the repeating decimal produced by 1/6.
35. Find the repeating decimal produced by 1/11.
36. Find the repeating decimal produced by 2/13.
37. Find the repeating decimal produced by 77/123.
38. Find the repeating decimal produced by 2/11.
39. Find the repeating decimal produced by 12/123.

Irrational numbers and non-repeating decimals Use a calculator.

40. Find the non-repeating decimal produced by square root of 3.
41. Find the non-repeating decimal produced by $\pi/2$.
42. Find the non-repeating decimal produced by log 2
43. Find the non-repeating decimal produced by ln 2
44. Find the non-repeating decimal produced by e^2
45. Explain why the ratio of two integers cannot be irrational.

Decimal addition and subtraction Find the sums and differences.

46.	47.	48.	49.
$23.005+322.1$	$23.005-322.1$	$3007.1+1.3002$	$3007.1-1.3002$

50.	51.	52.	53.
$100.01+1.01$	$100.01-1.01$	$5.06+3.067$	$5.06-3.067$

54.	55.
$90034.002+100.6$	$90034.002-100.6$

Decimal multiplication Find the products.

56.	57.	58.	59.	60.
23.005×322.1	3007.1×1.3002	100.01×1.01	5.06×3.067	90034.002×100.6

Decimal division Find the quotients

61 $23.005\div322.1$	62 $3007.1\div1.3002$	63 $100.01\div1.01$

64 $5.06\div3.067$	65 $90034.002\div100.6$

4 Algebraic Operations

Algebra is a branch of mathematics which processes the relations and properties of numbers by means of letters, signs of operations, and other symbols. Algebra is probably the reader's introduction to abstraction where the general, such as the letters x, are in use as opposed to the particular such as specific numbers $n= 5$ or 12.34 or -7 or -1008000.34. Here is the guiding principle of algebraic operations.

> The equality is not upset if the *same* change is made to both sides of =.

For example to solve for x in equation 1 one needs to isolate x on one side of =. If one subtracts 3 from both sides of equation 2, then x is isolated.

(1) $x+3=5$ $\rightarrow$ $x+3-3=5-3$ *subtract 3 from both sides of* =
to get $x=2$

Again, to solve for x in equation 2 one needs to isolate x on one side of =. If one adds 3 to both sides of equation 2, then x is isolated.

(2) $x-3=5$ $\rightarrow$ $x-3+3=5+3$ *add 3 to both sides of* = *to get* $x=8$

The *sign* of a term is changed when it is moved to the other side of =.

Addition and multiplication are the fundamental operations. Subtraction and division are the inverse operations of addition and multiplication respectively. Multiplication is connected to addition by the distributive law $x\times(y+z)=x\times y+x\times z$. The distributive law *expands* expressions.

Solving for x in equation 3 requires several steps. Multiply both sides of = by 4. Apply the distributive law. Then add 12 to both sides of =.

(3) $\dfrac{x}{4}-3=5$ $\rightarrow$ $4\times\left(\dfrac{x}{4}-3\right)=4\times5$ *multiply both sides of* = *by 4*

to get $4\times\dfrac{x}{4}-4\times3=4\times5$ $\rightarrow$ $x-12=20$ $\rightarrow$ $x=32$

check $\dfrac{32}{4}-3=8-3=5$ *qed*

This time divide both sides of = by 7 to solve for x.

(4) $7x - 3 = 5$ $\rightarrow$ $\frac{1}{7} \times (7x - 3) = \frac{1}{7} \times 5$ $\quad$ *divide both sides of = by 7*

to get $\frac{1}{7} \times 7x - \frac{1}{7} \times 3 = \frac{1}{7} \times 5$ $\rightarrow$ $x - \frac{3}{7} = \frac{5}{7}$ $\rightarrow$ $x = \frac{8}{7}$

check $7\frac{8}{7} - 3 = 8 - 3 = 5$ *qed*

The next example does not solve for variable x. The object is expanding (multiplying out) to get a sum of terms. All terms have coefficient 1 or -1. Making coefficients 1 explicit is always a useful tactic.

(5) $(x - y)(-x + y)$

$\quad = x(-x + y) - y(-x + y)$ $\quad$ *apply distributive law*

$\quad = -x^2 + xy + yx - y^2$ $\quad$ *apply distributive law again*

$\quad = -x^2 + 2xy - y^2$ $\quad$ *add equal terms* $xy = yx$

$\quad = (-1)x^2 + (1)2xy + (-1)y^2$ $\quad$ *make 1 coefficients explicit*

$\quad = (-1)x^2 + (-1)(-1)2xy + (-1)y^2$ $\quad 1 = (-1)(-1)$

$\quad = (-1)[x^2 + (-1)2xy + y^2]$ $\quad$ *factor out* -1, *a reverse distribution*

$\quad = -[x^2 - 2xy + y^2]$ $\quad\quad -= (-1)$

Expand this.

(6) $(x - y)^2 = (x - y)(x - y)$

$\quad\quad = x(x - y) - y(x - y)$ $\quad$ *apply distributive law*

$\quad\quad = x^2 - xy - yx + y^2$ $\quad$ *apply distributive law again*

$\quad\quad = x^2 - 2xy + y^2$ $\quad$ *add equal terms*

A closer look at part of equation 6.

(7) $-y(x - y) = (-y) \times (x) + (-y) \times (-y)$ *apply distributive law*

$\quad -y(x - y) = (-1)yx + (-1)(-1)yy$ $\quad$ *make coefficients 1 explicit*

$\quad -y(x - y) = -yx + y^2$

common error $\quad -y(x - y) = -yx - y^2$

Problem 401 Show that $x^2 - y^2 = (x - y)(x + y)$

Problem 402 Show that $x^3 - y^3 = (x - y)(x^2 + xy + y^2)$

Algebra

Word problems are used as means to learn how to translate text into equations. Problems such as this one.

A number x is tripled. Then 11 is added so that the sum equals 32. Solve for x.

(8) $3x+11=32 \rightarrow 3x+11-11=32-11 \rightarrow 3x=21 \rightarrow \frac{1}{3}x=\frac{1}{3}21 \rightarrow x=7$

Algebraic Operations on an Equation. Analysis of a problem produces an equation. Usually the equation needs to be recast into some other form. Recasting is done by algebraic operations such as addition, multiplication, subtraction, and division.

Convert $T_F = \frac{9}{5}T_C + 32$ *into* $T_C = \frac{5}{9}(T_F - 32)$

(1) *subtract 32 from both sides of* $=$ $\rightarrow$ $T_F - 32 = \frac{9}{5}T_C + 32 - 32$

(2) *clearly* $32 - 32 = 0$ *so that* $\rightarrow$ $T_F - 32 = \frac{9}{5}T_C$

(3) *in step 2 we moved 32 to the other side of* $=$ *and changed its sign*

(4) *multiply both sides of* $=$ *by* $\frac{5}{9}$ $\rightarrow$ $\frac{5}{9} \times (T_F - 32) = \frac{5}{9} \times \frac{9}{5}T_C$

(5) *notice addition of parens, because* $\frac{5}{9}$ *multiplies ALL terms*

(6) *clearly* $\frac{5}{9} \times \frac{9}{5} = 1$ *so that* $\rightarrow$ $T_C = \frac{5}{9}(T_F - 32)$ *qed*

(7) *in other words invert* $\frac{9}{5}$ *to* $\frac{5}{9}$ *in (2) and move to the other side*

Applying Geometry Algebraic problems are simplified and clarified by interpreting expressions and equations geometrically by representing lines, curves, and surfaces in a coordinate system.

Rene Descartes, a French mathematician, published in 1637 the first systematic work on merging the concepts of algebra and a geometric coordinate system now known as the Cartesian coordinate system.

The number line was introduced in the study of Arithmetic (Figure 401).

Figure 401 Part of a Number Line

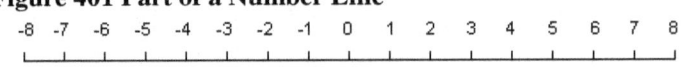

The number line is constructed by marking off equal lengths along the line. Each mark on the number line is assigned a number. Assign 0 to any mark. Next, assign 1 to the first mark to the right of zero. Then the *distance* from 0 to 1 *represents* 1 unit of length. Subsequent marks to the right add 1 unit to the distance. Label subsequent marks 2, 3, 4, and so forth. Repeat for negative numbers to the left.

The *distance* from 0 to 1 *represents* 1. The distance from 1 to 2 represents 1 more. Watch this. The distance from -3 to -2 also represents 1 more. The distance from 2 to 1 represents 1 less. The distance from -1 to -2 also represents 1 less. Move to the right to increase value. Move to the left to decrease value.

If a second number line is rotated by 90 degrees and superimposed on the horizontal number line at coordinate (0,0) an x, y coordinate system is established (Figure 402).

An elementary example of algebra in action is the derivation of equation (9a) of *any* straight line where letters x and y represent variables (the coordinates of the line), and letters c and d represent constants. The key word here is *any*. Variables represent any numbers, whereas constants represent specific numbers that do not change. Equation 9a is *one* form of the general straight line equation. Equations 9b and 9c are two specific examples.

$(9a)$ $y = cx + d$ $(9b)$ $y = -3x + 7$ $(9c)$ $y = 5.2x - 3.9$

Algebra

Figure 402 Cartesian Coordinate System

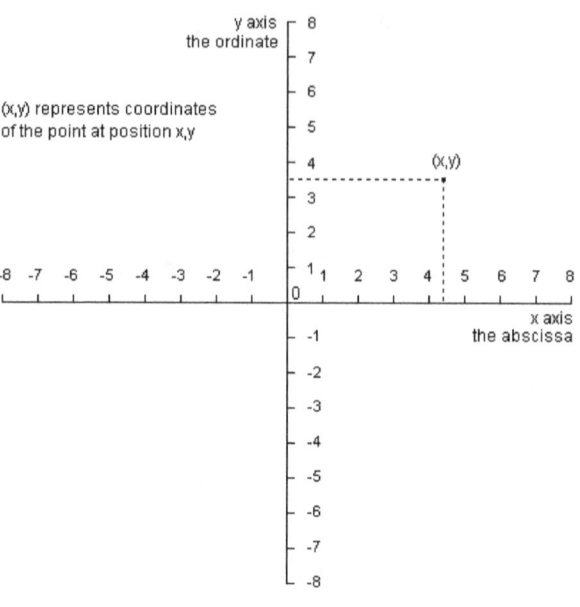

A line is drawn in the x, y coordinate system (Figure 403). Four points are marked on the line.

(x,y)=(c,0)   x axis intercept
(x,y)=(x_1,y_1)   a specific point
(x,y)=(0,b)   y axis intercept
(x,y)=(x,y)   any point x, y on the line

A walk from point (x, y) to point (x_1, y_1) causes changes in x and y, which are described as increments Δx and Δy (delta x and delta y). Δ is *the universal mathematics symbol for change in value.*

The slope *m* of a line is a measure of how steep a line is. A horizontal line has slope equal to zero ($\Delta y=0$). A slope m=1 means every step in x is matched by an equal step in y ($\Delta y = \Delta x$). The equation for slope m is defined as the ratio $\Delta y/\Delta x$.

(10) $m = \dfrac{increment\ in\ y}{increment\ in\ x} = \dfrac{\Delta y}{\Delta x} = \dfrac{y - y_1}{x - x_1}$ *created by points* (x, y) *and* (x_1, y_1)

As in arithmetic the four fundamental operations of algebra are addition, multiplication, subtraction and division.

Figure 403 Line in a Cartesian Coordinate System

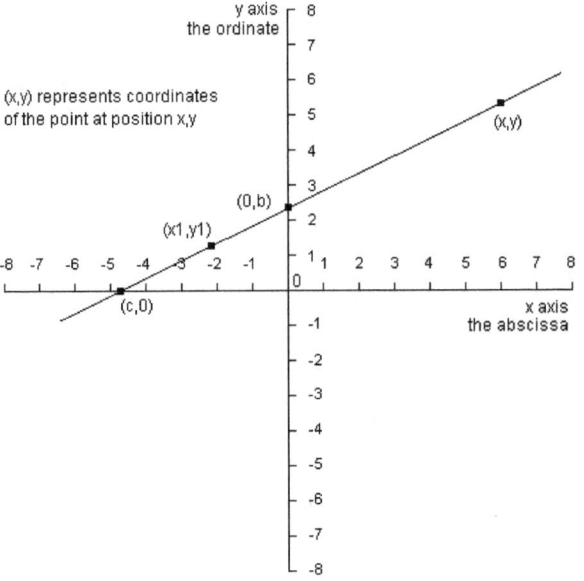

An equation equates expression A to expression B so that A=B.

Algebraic operations produce the equation of the line (Figure 403). Multiply both sides of = in equation 10 by $(x-x_1)$.

(11) $(x-x_1)m = \dfrac{y-y_1}{x-x_1}(x-x_1) \;\rightarrow\; (x-x_1)m = (y-y_1)$

Add y_1 to both sides of =.

(12) $(x-x_1)m + y_1 = y - y_1 + y_1 \qquad \rightarrow \qquad (x-x_1)m + y_1 = y$

(13) $y = m(x-x_1) + y_1$

Substitute point (0, b) for point (x_1, y_1) in equation 13. I.e. substitute 0 for x_1 and b for y_1.

(14) $y = mx + b \qquad$ *the slope intercept form of a line equation*

Algebra

Calculate the value of slope m. In equation 14 substitute point (c,0) for point (x,y). I.e. substitute c for x and 0 for y.

(15a) $0 = mc + b$

(15b) *subtract b from both sides of* $=$ $\rightarrow$ $0 - b = mc + b - b$ $\rightarrow$ $-b = mc$

(15c) *divide both sides of* $=$ *by c* $\rightarrow$ $m = -b\dfrac{1}{c} = -\dfrac{b}{c}$

Note: c is a negative number, which makes m a positive number

As an alternative define slope m by using x axis and y axis intercept points (c,0) and (0,b).

(16) $m = \dfrac{\Delta y}{\Delta x} = \dfrac{b-0}{0-c} = \dfrac{b}{-c}$ *positive slope of the line*

Yet another way is to let any point (x,y) and x axis intercept point (c,0) define the line equation's intercept form (equation 17i).

(17a) $m = \dfrac{b}{-c} = \dfrac{\Delta y}{\Delta x} = \dfrac{y-0}{x-c}$

(17b) *multiply both sides of* $=$ *by x−c* $\rightarrow$ $\dfrac{b}{-c}(x-c) = \dfrac{y-0}{x-c}(x-c)$

(17c) *cancel* $(x-c)/(x-c) = 1$, *and multiply by* $-c$ $\rightarrow$ $\dfrac{b}{-c}(x-c)(-c) = y(-c)$

(17d) *cancel* $(-c)/(-c) = 1$, *distribute b* $\rightarrow$ $bx - bc = -yc$

(17e) *subtract bx from both sides of* $=$ $\rightarrow$ $bx - bc - bx = -yc - bx$

(17f) *so that* $-bc = -yc - bx$

(17g) *multiply both sides of* $=$ *by* -1 $\rightarrow$ $bc = yc + bx$

(17h) *divide both sides of* $=$ *by bc* $\rightarrow$ $\dfrac{bc}{bc} = \dfrac{yc}{bc} + \dfrac{bx}{bc}$

(17i) *cancel* $\dfrac{bc}{bc} = 1$, $\dfrac{c}{c} = 1$, $\dfrac{b}{b} = 1$ $\rightarrow$ $1 = \dfrac{y}{b} + \dfrac{x}{c}$ *intercept form*

Looking at Figure 403, clearly the slope of the line m is positive. Slope m is positive, because numbers b and −c, repeat −c, are both positive.

There was no problem using (any number) c while developing equations. The fact that c is negative did not matter, which emphasizes what algebra is all about.

Temperature Scales

One application of the straight line is the relationship of the Fahrenheit temperature scale T_F to the Centigrade temperature scale T_C.

Water freezes at $32°$ and boils at $212°$ on the Fahrenheit temperature scale T_F, whereas it freezes at $0°$ and boils at $100°$ on the Centigrade temperature scale T_C. The slope of the line is the ratio of the two temperature increments $212-32$ and $100-0$. Designate T_F as the y axis and T_C as the x axis.

(1) $\quad m = \dfrac{increment\ in\ T_F}{increment\ in\ T_C} = \dfrac{212-32}{100-0} = \dfrac{180}{100} = \dfrac{9}{5}$

Any point (T_F, T_C) and T_F axis intercept point $(32,0)$ define the temperature equation's slope.

(2a) $\quad m = \dfrac{\Delta T_F}{\Delta T_C} \quad \rightarrow \quad \dfrac{9}{5} = \dfrac{T_F - 32}{T_C - 0}$

(2b) $\quad$ multiply both sides of $=$ by $T_C - 0 \quad \rightarrow \quad \dfrac{9}{5}(T_C - 0) = \dfrac{T_F - 32}{T_C - 0}(T_C - 0)$

(2c) $\quad$ cancel $(T_C - 0) \quad \rightarrow \quad \dfrac{9}{5}(T_C - 0) = T_F - 32$

(2d) $\quad$ add 32 to both sides of $= \quad \rightarrow \quad \dfrac{9}{5}(T_C - 0) + 32 = T_F - 32 + 32$

(2e) $\quad$ cancel 32, distribute $\dfrac{9}{5} \quad \rightarrow \quad T_F = \dfrac{9}{5}T_C + 32$

Problem 403 Show that $x^3 + y^3 = (x+y)(x^2 - xy + y^2)$

Problem 404 Show that $x^3 + y^3 = x^3 + 3x^2 y + 3xy^2 + y^3$

Problem 405 Show that $(x+y+z)^2 = x^2 + y^2 + z^2 + 2yz + 22zx + 2xy$

Problem 406 Show that

if $\dfrac{a}{b} = \dfrac{c}{d}$ and $k = \dfrac{a}{b} = \dfrac{c}{d}$ and $k \neq 0$, then $\dfrac{a+b}{a-b} = \dfrac{c+d}{c-d}$

Problem 407 Plot points $(q, 2q)$ where $q=0, \pm1, \pm2, \pm3, \pm4$

Electrical Feedback Amplifier

Analysis of an electrical feedback amplifier (Figure 801) produces three equations 1.

Figure 801 Feedback Amplifier

(1a) $v_3 = \mu v_2$

(1b) $v_2 = v_1 + v_4$

(1c) $v_4 = \beta v_3$

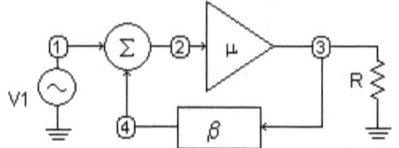

What follows is a typical sequence of many different operations such as replacement, distribution, subtraction, factoring, and division.

Solve for v_3/v_1. First v_2 is eliminated by substitution (replacement). Second v_4 is eliminated producing an equation in v_3 and v_1. Then solve for v_3/v_1.

(2a)
$$v_3 = \mu v_2$$
$$v_3 = \mu(v_1 + v_4) \qquad \text{replace } v_2 \text{ with } v_1 + v_4 \text{ (1b)}$$
$$v_3 = \mu(v_1 + \beta v_3) \qquad \text{replace } v_4 \text{ with } \beta v_3 \text{ (1c)}$$
$$v_3 = \mu v_1 + \mu \beta v_3 \qquad \text{use distributive law}$$

(2b) $v_3 - \mu \beta v_3 = \mu v_1 + \mu \beta v_3 - \mu \beta v_3 \qquad$ subtract $\mu \beta v_3$ from both sides of $=$

$$v_3 - \mu \beta v_3 = \mu v_1$$
$$v_3(1 - \mu \beta) = \mu v_1 \qquad \text{factor out } v_3$$

(2c)
$$v_3 = \frac{\mu v_1}{1 - \mu \beta} \qquad \text{divide both sides of } = \text{ by } 1 - \mu \beta$$

$$\frac{v_3}{v_1} = \frac{\mu}{1 - \mu \beta} \qquad \text{divide both sides of } = \text{ by } v_1$$

Problem 408 Plot points (x, y) where x= 0, ±1, ±2 and y=x²+1.

Problem 409 Plot points (x, y) where x ≤ 8, and y = −2.

Problem 410 Plot points (x, y) where x = 3, and y = −2 to y = +5.

Problem 411 Plot points (q, 2q) where q=0, ±1, ±2, ±3, ±4

Problem 412 Show that the equation for points (x, y) = (−4, −1), (−2, 1), (0, 3), (2, 5) is y = x+3.

---

5 Polynomials

If x is a *variable*, then x can be *any* number real or complex. If a_j is a *constant*, then a_j can be *any specific* number real or complex. One can build a *polynomial* as follows from degree 0 to degree n, where the *degree* equals the highest power of x.

(1a) $f_0(x) = a_0$

(1b) $f_1(x) = a_1 x + a_0$

(1c) $f_2(x) = a_2 x^2 + a_1 x + a_0$

and so forth until the mathematician's favorite degree n is reached

(1n) $f_n(x) = a_n x^n + a_{n-1} x^{n-1} + \cdots + a_2 x^2 + a_1 x + a_0$

Reminder: what is a power of x such as x^n when n is an integer?

(2) $x^n = x \times x \times x \times \cdots \times x \times x$ *(the product of n x's)*

Next build a polynomial in a way that emphasizes *roots*. A root r of a polynomial is a value of x that makes the polynomial equal zero.

(3a) $f_0(x) = 1$

(3b) $f_1(x) = (x - r_1)$

(3c) $f_2(x) = (x - r_1)(x - r_2)$

and so forth until the mathematician's favorite degree n is reached

(3n) $f_n(x) = (x - r_1)(x - r_2) \cdots (x - r_{n-1})(x - r_n)$

Apply the distributive law to equation 3c.

(4) $f_2(x) = (x - r_1)(x - r_2)$

$\qquad = x(x - r_2) - r_1(x - r_2)$

$\qquad = x^2 - r_2 x - r_1 x + r_1 r_2$

$\qquad = x^2 - (r_2 + r_1)x + r_1 r_2$

Equate equations 1c and 4 in order to relate roots to coefficients.

(5a) 1c $f_2(x) = 4\ f_2(x)$

(5b) $a_2 x^2 + a_1 x + a_0 = x^2 - (r_2 + r_1)x + r_1 r_2$

(5c) $(a_2 - 1)x^2 + (a_1 + (r_2 + r_1))x + (a_0 - r_1 r_2) = 0$

(5d) *so that for any x* $a_2 = 1,\quad a_1 = -(r_2 + r_1),\quad a_0 = r_1 r_2$

Algebra

5.1 Operations

Polynomial addition and subtraction Addition is implemented by grouping of powers of x (equation 6b), and adding the grouped coefficients (equation 6d). Subtraction is implemented by grouping of powers of x and subtracting the coefficients. Another way to add polynomials is to use arithmetic format and add each column (equation 6e). There are no carries.

add the polynomials

(6a) $\quad (2x^5 - 7x^4 + x^3 + 3x^2 - 2x + 1) + (-x^5 + 3x^4 - 16x^2 - 4x + 5)$

group by powers of x

(6b) $\quad (2x^5 - x^5) + (-7x^4 + 3x^4) + (x^3) + (3x^2 - 16x^2) + (-2x - 4x) + (1 + 5)$

apply the distributive law in reverse, factor out the powers of x

(6c) $\quad (2-1)x^5 + (-7+3)x^4 + (1+0)x^3 + (3-16)x^2 + (-2-4)x + (1+5)$

add the coefficients

(6d) $\quad x^5 - 4x^4 + x^3 - 13x^2 - 6x + 6$

redo the addition in Arithmetic format, add term by term

$$
\begin{array}{l}
 2x^5 - 7x^4 + x^3 + 3x^2 - 2x + 1 \\
(6e)\ \underline{-x^5 + 3x^4 - 16x^2 - 4x + 5} \\
 x^5 - 4x^4 + x^3 - 13x^2 - 6x + 6
\end{array}
$$

Polynomial multiplication in arithmetic format applies the distributive law in the same way as in arithmetic where each term now is a power of x instead of a number. The last step adds the rows of each column with no carries.

(7a)

$$
\begin{array}{llll}
 & x^5 - x^3 + 2x + 1 & f(x) \\
\underline{\times \qquad\qquad\qquad 2x^3 + 5x + 1} & g(x) \\
\qquad x^5 \qquad -x^3 \qquad +2x + 1 & 1 \times f(x) \\
\quad 5x^6 \quad -5x^4 \qquad +10x^2 + 5x & 5x \times f(x) \\
\underline{2x^8 - 2x^6 \quad +4x^4 + 2x^3} & 2x^3 \times f(x) \\
2x^8 + 3x^6 + x^5 - x^4 + x^3 + 10x^2 + 7x + 1 & sum\ of\ 3\ rows
\end{array}
$$

(7b) *check* — *sum of 3 rows* $S = 1 \times f(x) + 5x \times f(x) + 2x^3 \times f(x)$

(7c) *factor out* $f(x)$ $\quad S = f(x)(1 + 5x + 2x^3) = f(x) \times g(x)$ *qed*

Polynomial multiplication using the distributive law directly.
Polynomials to multiply

$(8a)\quad (2x^3 + 5x + 1)(x^5 - x^3 + 2x + 1)$

apply the distributive law

$(8b)\quad 2x^3(x^5 - x^3 + 2x + 1) + (5x)(x^5 - x^3 + 2x + 1) + (1)(x^5 - x^3 + 2x + 1)$

multiply

$(8c)\quad (2x^8 - 2x^6 + 4x^4 + 2x^3) + (5x^6 - 5x^4 + 10x^2 + 5x) + (x^5 - x^3 + 2x + 1)$

gather coefficients of powers of x

$(8d)\quad (2)x^8 + (-2+5)x^6 + (1)x^5 + (4-5)x^4 + (2-1)x^3 + (10)x^2 + (5+2)x + 1$

add coefficients

$(8e)\quad 2x^8 + 3x^6 + x^5 - x^4 + x^3 + 10x^2 + 7x + 1$

Polynomial Division Divide h(x) by p(x) to get quotient q(x) and remainder r(x). Use the polynomials in equations 8a and 8e. The key is that quotient term x^5 times p(x) produces $2x^8$ as the first term, which cancels $2x^8$ in h(x). The value of the quotient terms x^5, $-x^3$, 0, 2x, 1 is determined by the requirement to cancel the next term in h(x).

$(9)\quad \dfrac{h(x)}{p(x)} = q(x) + \dfrac{r(x)}{p(x)} \quad \rightarrow \quad p(x)\overline{)h(x)} \quad where\, r(x)\, appears\, in\, the\, last\, step$

(10)

$$
\begin{array}{r}
x^5 - \ x^3 + 0 + 2x + 1 \qquad\qquad quotient\\
2x^3 + 5x + 1\overline{)2x^8 + 3x^6 + x^5 - x^4 + x^3 + 10x^2 + 7x\ +1} \quad h(x)/p(x)\\
2x^8 + 5x^6 + x^5 \qquad\qquad\qquad\qquad x^5 p(x)\\
0\ -2x^6 + 0\ -x^4 + x^3 + 10x^2 + 7x\ +1 \quad subtract\\
-2x^6 \qquad -5x^4 - x^3 \qquad\qquad\qquad -x^3 p(x)\\
0\ +\ 0 + 4x^4 + 2x^3 + 10x^2 + 7x + 1 \quad subtract\\
4x^4 \qquad +10x^2 + 2x \qquad 2xp(x)\\
0\ +2x^3 \qquad +5x + 1 \quad subtract\\
2x^3 \qquad +5x + 1 \quad 1\times p(x)\\
0 \quad subtract
\end{array}
$$

Problem 501 Divide $f(x) = x^3 - 3x^2 + 4$ by $x - 2$ to get quotient $x^2 - x - 2$ and remainder 0.

Problem 502 Divide $f(x) = 3x^3 - 13x^2 + 13x - 3$ by $x - 1$ to get quotient $3x^2 - 10x + 3$ and remainder 0.

5.2 The Remainder Theorem

In arithmetic division of x by y produced a quotient q and a remainder r where $x = qy + r$. The quotient q equals the number of y's in x and the remainder r equals a partial y that is left over. For example there are 196 y's (23's) in 4521 and a partial y of 13 is left over.

(11a) $\dfrac{dividend}{divisor} = quotient + \dfrac{remainder}{divisor} \rightarrow \dfrac{x}{y} = q + \dfrac{r}{y} \rightarrow \dfrac{4521}{23} = 196 + \dfrac{13}{23}$

(11b) $x = qy + r \rightarrow 4521 = 196 \times 23 + 13$

One way to determine whether or not a number is a root of a polynomial is to use the division process.

For example to find out whether or not 2 is a root of polynomial h(x) divide h(x) by (x–2) until no x appears in the remainder. I.e. the *degree* of remainder r(x) is 0, which is less than the degree 1 of the divisor. Then the ratio r(x)/(x–2) is a proper fraction that cannot be reduced, which is why the process terminates. The fraction 32/(x–2) cannot be reduced.

(12)

$$
\begin{array}{r}
x^2 + 7x + 10 \\
x - 2 \overline{)\, x^3 + 5x^2 - 4x + 12} \\
\underline{x^3 - 2x^2} \\
7x^2 - 4x \\
\underline{7x^2 - 14x} \\
10x + 12 \\
\underline{10x - 20} \\
32
\end{array}
$$

An equation for h(x) that is valid for all x is

(13) $h(x) = p(x)q(x) + r(x)$

$h(x) = (x-2)(x^2 + 7x + 10) + 32$ *and* $h(2) = 32 = r(2)$

Use the Remainder Theorem to avoid the division process (equation 12).

Remainder Theorem 1 *The remainder r(b) when the polynomial*
$$f(x) = a_n x^n + a_{n-1} x^{n-1} + \cdots + a_2 x^2 + a_1 x + a_0 \text{ is divided by } (x-b) \text{ is}$$
$$r(b) = f(b) = a_n b^n + a_{n-1} b^{n-1} + \cdots + a_2 b^2 + a_1 b + a_0$$

Proof Let $q(x)$ be the quotient, a polynomial of degree n−1 in x, and let $r(x)$ be the remainder when $f(x)$ is divided by $(x-b)$. Then

$$f(x) = (x-b)q(x) + r(x)$$

Since the equation is true for all values of x it is true when x=b, and so

$$f(b) = (b-b)q(b) + r(b) \quad \rightarrow \quad f(b) = r(b)$$

Remainder Theorem 2 *If f(b)=0 then the polynomial*

$$f(x) = a_n x^n + a_{n-1} x^{n-1} + \cdots + a_2 x^2 + a_1 x + a_0$$

has (x−b) as a factor. Conversely if (x−b) is a factor of f(x), then f(b)=0.

Proof By Remainder Theorem 1 when $f(x)$ is divided by $(x-b)$ the remainder is $f(b)$.

$$f(b) = a_n b^n + a_{n-1} b^{n-1} + \cdots + a_2 b^2 + a_1 b + a_0$$

If $f(b)=0$ the remainder $r(b)=0$, then $(x-b)$ divides $f(x)$ exactly so that $(x-b)$ is a *factor* of $f(x)$.

Conversely if $(x-b)$ is a factor of $f(x)$, then $f(b)=0$, and the remainder $r(x)=r(b)$ equals zero.

$$f(x) = (x-b)q(x) + r(x)$$
$$f(b) = (b-b)q(b) + r(b) = r(b) \quad \rightarrow \quad if \ f(b) = 0, \ then \ r(b) = 0$$

5.3 Factors of Polynomials and the Remainder Theorem

Factoring is the reverse of the distributive operation. Going from left to right in $(x-2)(x+3) = x^2 + x - 6$ is expanding. Going from right to left in $(x-2)(x+3) = x^2 + x - 6$ is factoring.

Find the roots (factors) of $h(x) = 3x^3 - 13x^2 + 13x - 3$. The constant term 3 equals the product of the roots (equation 5d page 49). Therefore the possible roots of h(x) must include ± 1, ± 3, which are the factors of 3.

(14a) $h(1) = 3 - 13 + 13 - 3 = 0$

(14b) $h(-1) = -3 - 13 - 13 - 3 \neq 0$

(14c) $h(3) = 3 \cdot 27 - 13 \cdot 9 + 13 \cdot 3 - 3 = 81 - 117 + 39 - 3 = 0$

(14d) $h(-3) = 3 \cdot (-27) - 13 \cdot 9 + 13 \cdot (-3) - 3 = -81 - 117 - 39 - 3 \neq 0$

(14e) $h(x) = (x-1)(x-3)g(x) = (x^2 - 4x + 3)g(x)$

Find g(x)=h(x)/(x²–4x+3). One way is to divide.

(14f)
$$
\begin{array}{r}
3x - 1 \\
x^2 - 4x + 3 \overline{)\,3x^3 - 13x^2 + 13x - 3} \\
\underline{3x^3 - 12x^2 + 9x} \\
-x^2 + 4x - 3 \\
\underline{-x^2 + 4x - 3} \\
0
\end{array}
$$

(14g) $g(x) = 3x - 1$

(14h) $h(x) = (x-1)(x-3)(3x-1)$

For example prove by means of the remainder theorem that *(2x–1)* is a factor of $f(x) = 2x^6 - 3x^5 + x^4 - 2x^2 + 3x - 1$

(16a) $(2x - 1) = 2\left(x - \frac{1}{2}\right)$ *the root is* $\frac{1}{2}$

(16b) $f(\frac{1}{2}) = 2\frac{1}{2}^6 - 3\frac{1}{2}^5 + \frac{1}{2}^4 - 2\frac{1}{2}^2 + 3\frac{1}{2} - 1 = \frac{1}{32} - \frac{3}{32} + \frac{1}{16} - \frac{1}{2} + \frac{3}{2} - 1 = 0$

Harder. Find the factors of $f(x, y) = 3x^3 - 22x^2 y + 43xy^2 - 12y^3$.

Think of this as a third degree polynomial in x whose coefficients of x are $3, -22y, 43y^2, -12y^3$.

Apply *Remainder Theorem 2*. Note that $-12y^3$ in f(x) is the product of the roots, which has the factors $-12, \pm4, \pm3$, y, y, y and combinations thereof.

(17a) $f(y, y) = 3y^3 - 22y^2 y + 43yy^2 - 12y^3 \neq 0$

(17b) $f(3y, y) = 3 \cdot 3^3 y^3 - 22 \cdot 3^2 y^3 + 43 \cdot 3y \cdot y^2 - 12y^3$

$\qquad\qquad = (81 - 198 + 129 - 12)y^3 = 0 \quad and \; so \, (x - 3y) \; is \; a \; factor$

Problems 503 and 504 complete the factoring process.

Problem 503 Divide by (x−3y) to get $f(x, y) = (x - 3y)(3x^2 - 13yx + 4y^2)$.

Problem 504 Show that $f(x, y) = (x - 3y)(3x - y)(x - 4y)$

Algebra

5.4 The Zeros of a Polynomial

We offer without proof the

Fundamental Theorem of Algebra Every integral rational equation in one variable x has at least one zero (root).

Corollary Every polynomial of degree n can be decomposed into n linear factors.

Proof Let
$$f(x) = a_n x^n + a_{n-1} x^{n-1} + \cdots + a_2 x^2 + a_1 x + a_0 \qquad (a_n \neq 0)$$

By the *Fundamental Theorem* $f(x)=0$ has at least one zero z_1 so that
$$f(x) = (x - z_1) q_1(x).$$

By the *Fundamental Theorem* $q_1(x)=0$ has at least one zero z_2 so that
$$q_1(x) = (x - z_2) q_2(x) \quad and \ so \quad f(x) = (x - z_1)(x - z_2) q_2(x).$$

Repeating the process n times
$$f(x) = a_n (x - z_1)(x - z_2) \cdots (x - z_n) \qquad where \ the \ z_k \ are \ the \ n \ zeros \ of \ f(x)$$

Theorem Every integral rational equation of degree n in one variable x has no more than n zeros.

Proof Let m, n be any numbers, then from the *Corollary*
$$f(x) = a_n (x - z_1)(x - z_2) \cdots (x - z_n)$$

Clearly $f(x)$ cannot equal zero when x has any value m distinct from the n zeros z_k.

Note: The zeros need not all be real and distinct. For example if $z = z_3 = z_4 = z_9$, then z is a *multiple zero of order 3*. Nor do they have to be real. If a zero is a complex number $z=a+ib$, then some other zero has to be the complex conjugate $z=a-ib$ when the coefficients of $f(x)$ are real. In that case complex zeros always occur in conjugate pairs, and may be of any order. In any case the number of zeros cannot exceed n, the degree of $f(x)$.

5.5 Newton's Method for finding Zeros

Let $f(x)$ denote any polynomial in x and let $f'(x)$ denote its differential coefficient. Then

(18) $\dfrac{f(a+h)-f(a)}{h} \to f'(a)$ $\qquad$ as $h \to 0$

It follows that, when h is small

(19a) $f(a+h)-f(a) \approx hf'(a)$

(19b) $f(a+h) \approx f(a)+hf'(a)=0 \ \Rightarrow \ h = -\dfrac{f(a)}{f'(a)}$

We apply equation 19b to the problem of finding the numerical value of a zero of a polynomial **If $f(x)=x^n$, then $f'(x)=nx^{n-1}$**.

(20a) let $f(x) \equiv x^3 - 3x^2 + 4x - 3$

(20b) so that $f'(x) = 3x^2 - 6x + 4 + 0$

A table of values shows that there is a zero between x=1 and x=2, because f(x) changed sign from -1 to 1.

(21)
$x =$	0	1	2	3
$f(x) =$	-3	-1	1	9

Since the x values are 1 and 2 let the first estimate for the zero's value be a=1.5. Then from 19b

(22a) $f(1.5+h_1) \approx f(1.5)+h_1 f'(1.5) = -\frac{3}{8}+h_1\frac{7}{4}=0$

(22b) $h_1 = \frac{3}{8}\frac{4}{7} = \frac{12}{56} \approx 0.2 \ \Rightarrow \ $ next estimate of zero $= 1.7$

(22c) $f(1.7+h_2) \approx f(1.7)+h_2 f'(1.7) = 0.043+h_2 2.47 = 0$

(22d) $h_2 = -\frac{0.043}{2.47} \approx -0.02 \ \Rightarrow \ $ 3rd estimate of zero $= 1.68$ and so forth

---

Use Newton's method to find one *real* root r.

Problem 505 $x^7 - 5x + 3 = 0$ $\quad$ (r=0.6060)

Problem 506 $x^5 - 3x^2 - 8 = 0$ $\quad$ (r=1.77075)

Problem 507 $x^3 + 7x - 3 = 0$ $\quad$ (r=0.4180)

Problem 508 $70x^3 - 81x^2 - 100x + 96 = 0$ $\quad$ (r=0.8000)

Problem 509 $14x^4 + 23x^3 - 16x^2 + 23x - 30 = 0$ $\quad$ (0.860)

---

6 Polynomial Equations

An *equality* is a statement that two algebraic expressions are equal. The two expressions are referred to as members or sides of the equality. There are two kinds of equalities - identities and equations.

We have an *identical equality*, or simply an *identity*, if the two members of the equality are equal for *all values* of the symbols for which the members are defined. For example use the distributive law to show that equations 1a and 1b are identities.

$$(1a) \quad x^2 - y^2 = (x+y)(x-y) \qquad (1b) \quad (2x-1)^2 + 4x = 4x^2 + 1$$

We have an *equation*, if the two members of the equality are equal only for certain particular values of the symbols for which the members are defined. For example

$$(2a) \quad x - 3 = 2 \ (only \ for \ x = 5) \qquad (2b) \quad x^2 + 2 = 3x \ (only \ for \ x = 1 \ and \ 2)$$

An equation in one unknown x is an algebraic statement expressing a condition that the variable x must satisfy.

An equation in two unknowns x, y is an algebraic statement relating x and y that does not restrict either x *or* y to specific value(s). For example calculate y when x takes integer values 1, 2, 3, ...

$$(3a) \quad y = 5x + 3 \qquad (3b) \quad y = 5x^3 + 2x + 17$$

Degree of a term Count the letters to find the degree of a term with respect to a specific letter. For example the term $xxyyyz$ is a fourth degree term in y and z, second degree in x, third degree in y, and first degree in z. The degree of an equation with respect to any letter is the highest degree of that letter.

Solving an equation To solve an equation is to find *all* of its solutions. Any value of one unknown, or set of values of all unknowns, that makes both sides equal is a solution of the equation. The equation is *satisfied*.

Allowable operations on an equation
Adding the same number or expression to, or subtracting the same number or expression from, *both* sides.

Multiplying or dividing *both* sides by the same number or expression, provided the divider is not zero.

6.1 The Roots of an Equation

Given constants a_n, a_{n-1}, ... , a_0 ($a_n \neq 0$) let x be a variable. Then

(4) $\quad f(x) = a_n x^n + a_{n-1} x^{n-1} + \cdots + a_2 x^2 + a_1 x + a_0$

is a *polynomial* of degree n in x. The values of x for which f(x)=0 are referred to as the zeros, or roots, of f(x).

By the *Fundamental Theorem of Algebra* if b is a root then (x–b) is a factor of f(x). If $(x-b)^2$ is not a factor of f(x), then (x–b) is a *simple root* or a *non-repeated root*. If $(x-b)^3$ is not a factor of f(x), then $(x-b)^2$ may be a *double root*, and so forth.

If $(x-b)^2$ is a double root, then we may write

(5) $\quad f(x) = a_n (x-b)^2 g(x) \quad where$

$\quad g(x) = (x^{n-2} + c_{n-3} x^{n-3} + \cdots + c_2 x^2 + c_1 x + a_0)$

This nth degree polynomial with n roots has only n–1 *distinct* roots, because root b is counted twice.

For example this f(x) has 4 roots of which only 2 are distinct.

(6) $\quad f(x) = x^4 - 2x^2 + 1 = (x-1)^2 (x+1)^2 \quad has \ roots \ 1, \ 1, \ -1, \ -1$

Consequently it can be proved that every degree n polynomial can be expressed as the product of a constant and n linear factors (x–k) not necessarily distinct.

Algebra

6.2 Linear Equations

The general equation of the first degree in x and y is
(7) $ax + by + c = 0$

We can solve for x or y. The solution for y is as follows
(8a) $ax + by + c - (ax + c) = 0 - (ax + c)$ *subtract $(ax + c)$ from both sides*
(8b) $by = -ax - c$

(8c) $y = -\dfrac{a}{b}x - \dfrac{c}{b}$ *if $b \neq 0$, divide both sides by b*

On the other hand if a or b=0, then the equation reduces to one unknown, and the solution is straightforward.

(9) $ax + c = 0$ $(b = 0)$ $\rightarrow$ $x = -\dfrac{c}{a}$ *if $a \neq 0$, can divide both sides by a*

(10) $by + c = 0$ $(a = 0)$ $\rightarrow$ $y = -\dfrac{c}{b}$ *if $b \neq 0$, can divide both sides by b*

Algebraic solution of two equations   There are two basic ways to find solutions: by elimination and by determinants. Solution by elimination takes two forms: (1) by addition or subtraction, (2) by substitution. For example solve the equations $3x + 7y + 9 = 0$ *and* $4x + 5y - 1 = 0$ for x and y. (As a practical matter the following process can be applied to 3 equations.)

1 Solution by addition or subtraction.
(11) $12x + 28y + 36 = 0$ *multiply by 4*
       $-12x + 15y - 03 = 0$ *multiply by 3*
       $\overline{}$
               $13y + 39 = 0$ *solve to get $y = -3$*
$4x + 5y - 1 = 4x - 15 - 1 = 4x - 16 = 0$ *solve to get $x = 4$*
check $3x + 7y + 9 = 3(4) + 7(-3) + 9 = 0$ *qed*
check $4x + 5y - 1 = 4(4) + 5(-3) - 1 = 0$ *qed*

2 Solution by substitution.
(12) $x = \frac{1}{3}(-7y - 9)$ *first equation*

    $4\frac{1}{3}(-7y - 9) + 5y - 1 = 0$ *substitute for x in second equation*

    $-\frac{28}{3}y - 12 + 5y - 1 = -\frac{13}{3}y - 13 = 0$ *solve to get $y = -3$*

    $x = \frac{1}{3}(-7y - 9) = \frac{1}{3}(-7)(-3) - 3 = 7 - 3 = 4$ *solve to get $x = 4$ qed*

3 Solution by Determinants We manipulate two general simultaneous equations to show the origin of determinants.

(13a) $a_1 x + b_1 y + c_1 = 0$

(13b) $a_2 x + b_2 y + c_2 = 0$

(14) $b_2 a_1 x + b_2 b_1 y + b_2 c_1 = 0$ *multiply by* b_2

    $-b_1 a_2 x - b_1 b_2 y - b_1 c_2 = 0$ *multiply by* $-b_1$

$(b_2 a_1 - b_1 a_2)x + (b_2 c_1 - b_1 c_2) = 0$ *add, solve for* x

$x = -\dfrac{(b_2 c_1 - b_1 c_2)}{(a_1 b_2 - a_2 b_1)}$ *multiply by* a_1, a_2 *to get* $y = -\dfrac{(a_2 c_1 - a_1 c_2)}{(a_1 b_2 - a_2 b_1)}$

The values for x and y in (14) are examples of *determinants* that are written in (16) as determinants in the form of columns of coefficients. The symbol for a determinant is an $n \times n$ array of numbers bounded by bars (equation 15).

(15) $\begin{vmatrix} a_1 & b_1 \\ a_2 & b_2 \end{vmatrix} = a_1 b_2 - a_2 b_1$

From (13) $-c_1 = a_1 x + b_1 y$ *and* $-c_2 = a_2 x + b_2 y$

Then let $\Delta = a_1 b_2 - a_2 b_1$ *and*

(16) $x = \dfrac{\begin{vmatrix} c_1 & b_1 \\ c_2 & b_2 \end{vmatrix}}{\begin{vmatrix} a_1 & b_1 \\ a_2 & b_2 \end{vmatrix}} = -\dfrac{c_1 b_2 - c_2 b_1}{\Delta}$ $y = \dfrac{\begin{vmatrix} a_1 & c_1 \\ a_2 & c_2 \end{vmatrix}}{\begin{vmatrix} a_1 & b_1 \\ a_2 & b_2 \end{vmatrix}} = -\dfrac{a_1 c_2 - a_2 c_1}{\Delta}$

Substitute the numbers to solve for *x* and *y*.

(17a) $-c_1 = a_1 x + b_1 y$ $\rightarrow$ $-9 = 3x + 7y$

(17b) $-c_2 = a_2 x + b_2 y$ $\rightarrow$ $1 = 4x + 5y$

$\begin{vmatrix} c_1 & b_1 \\ c_2 & b_2 \end{vmatrix} = \begin{vmatrix} -9 & 7 \\ +1 & 5 \end{vmatrix} = -9 \times 5 - 7 \times 1 = -52$ $\begin{vmatrix} a_1 & c_1 \\ a_2 & c_2 \end{vmatrix} = \begin{vmatrix} 3 & -9 \\ 4 & +1 \end{vmatrix} = 3 \times 1 - (-9) \times 4 = 39$

Then $\Delta = a_1 b_2 - a_2 b_1 = 3 \times 5 - 7 \times 4 = -13$ *and*

(18) $x = \dfrac{\begin{vmatrix} c_1 & b_1 \\ c_2 & b_2 \end{vmatrix}}{\begin{vmatrix} a_1 & b_1 \\ a_2 & b_2 \end{vmatrix}} = -\dfrac{c_1 b_2 - c_2 b_1}{\Delta} = \dfrac{-52}{-13} = 4$ $y = \dfrac{\begin{vmatrix} a_1 & c_1 \\ a_2 & c_2 \end{vmatrix}}{\begin{vmatrix} a_1 & b_1 \\ a_2 & b_2 \end{vmatrix}} = -\dfrac{a_1 c_2 - a_2 c_1}{\Delta} = \dfrac{39}{-13} = -3$

6.3 Quadratic Equations

The general quadratic equation is a *polynomial* of *degree* 2 where a, b, c are numbers of any kind, and whose roots are r_1, r_2.

(19) $ax^2 + bx + c = 0$ $\quad (a \neq 0)$

(20) $ax^2 + bx + c = a(x - r_1)(x - r_2)$

Now

(21) $(x - r_1)(x - r_2) = x(x - r_2) - r_1(x - r_2) = x^2 - (r_2 + r_1)x + r_1 r_2$

and

(22a) *if* $ax^2 + bx + c = ax^2 - a(r_2 + r_1)x + a r_1 r_2$

      then equate coefficients

(22b) $b = -a(r_2 + r_1),$ $\qquad c = a r_1 r_2$

(22c) $r_2 + r_1 = -\dfrac{b}{a},$ $\qquad r_1 r_2 = \dfrac{c}{a}$

If the roots of the quadratic equation

$ax^2 + bx + c = 0\,(a \neq 0)$ *are* r_1, r_2 *then* $r_2 + r_1 = -\dfrac{b}{a},$ $\; r_1 r_2 = \dfrac{c}{a}$

Conversely

If r_1, r_2 *are the roots of* $x^2 + jx + k = 0$ *then* $j = -(r_2 + r_1),$ $\; k = r_1 r_2$

Solution by Completing the Square The roots of any quadratic equation can be found by a *completing the square* process, which creates a polynomial that is the square of another polynomial. In Chapter 1 we were reminded that *multiplication is connected to addition* by the distributive law $x(y + z) = xy + xz$, which is used to derive the following.

(23) $(x + a)(x + a) = x(x + a) + a(x + a) = x^2 + xa + ax + a^2$

      so that $(x + a)^2 = x^2 + 2ax + a^2$

let $m = 2a$ *(see eqn 23)* $\Rightarrow a = \dfrac{m}{2}$

Therefore to complete the square of $x^2 + mx$ *add* $a^2 = \left(\dfrac{m}{2}\right)^2 = \dfrac{m^2}{4}$

so that $x^2 + mx + \dfrac{m^2}{4} = (x + \dfrac{m}{2})^2$

Example

(24a) *find the roots of* $2x^2 - 6x - 5 = 0$

(24b) *add 5 to both sides* $2x^2 - 6x = 5$

(24c) *divide by 2* $\quad x^2 - 3x = \dfrac{5}{2}$

(24d) *add* $\left(\dfrac{-3}{2}\right)^2 = \dfrac{9}{4}$ *to both sides* $\rightarrow x^2 - 3x + \dfrac{9}{4} = \dfrac{5}{2} + \dfrac{9}{4} \rightarrow \left(x - \dfrac{3}{2}\right)^2 = \dfrac{19}{4}$

(24e) $x - \dfrac{3}{2} = \pm\dfrac{\sqrt{19}}{2} \quad \rightarrow \quad x = \dfrac{3}{2} \pm \dfrac{\sqrt{19}}{2}$

Example

(25a) *find the roots of* $4w^2 - 7 = -8w$

(25b) *add* $7 + 8w$ *to both sides* $4w^2 + 8w = 7$

(25c) *divide by 4* $\quad w^2 + 2w = \dfrac{7}{4}$

(25d) *add 1 to both sides* $\rightarrow w^2 + 2w + 1 = \dfrac{7}{4} + 1 \rightarrow (w+1)^2 = \dfrac{11}{4}$

(25e) $w + 1 = \pm\dfrac{\sqrt{11}}{2} \quad \rightarrow \quad w = -1 \pm \dfrac{\sqrt{11}}{2}$

Solve the pairs of linear equations algebraically, and check the results.

Problem 601 $2x - y + 7 = 0 \quad 3x + 4y - 6 = 0$ $\qquad$ (x,y=−2,3)

Problem 602 $2x - 3y = 10 \quad 2(x-10) - 3(1+2y) = 5 - 3x$ $\quad$ (x,y=8,2)

Problem 603 $3(2x-1) - 5(2y+1) = 0 \quad 5(3y-2) - 2(5x-5) + 15 = 0$
$\qquad\qquad\qquad\qquad\qquad\qquad\qquad$ (x,y=3,1)

Solve by completing the square

Problem 604 $3x^2 + 11x = 4$ $\qquad\qquad$ x=−11/6±√(169/36)

Problem 605 $x^2 + mx + n = 0$ $\qquad\qquad$ x=0.5(−m±√(m²-4n))

Problem 606 $px(r-q) = (px)^2 - qr$ $\qquad$ x=(−q/p, r/p)

Algebra

Solution by Quadratic Formula We express the roots as functions of coefficients a, b, c by applying the process of completing the square. To start transpose c and divide by a.

(26) $ax^2 + bx + c = 0 \rightarrow x^2 + \dfrac{b}{a}x = -\dfrac{c}{a}$

$\qquad$ next add $\left(\dfrac{b}{2a}\right)^2 = \dfrac{b^2}{4a^2}$ to both sides

$$x^2 + \dfrac{b}{a}x + \dfrac{b^2}{4a^2} = \dfrac{b^2}{4a^2} - \dfrac{c}{a} = \dfrac{b^2}{4a^2} - \dfrac{4ac}{4aa}$$

$$\left(x + \dfrac{b}{2a}\right)^2 = \dfrac{b^2 - 4ac}{4a^2} \qquad \rightarrow \qquad x + \dfrac{b}{2a} = \pm\dfrac{\sqrt{b^2 - 4ac}}{2a}$$

$$x_1, x_2 = -\dfrac{b}{2a} \pm \dfrac{\sqrt{b^2 - 4ac}}{2a}$$

Specific cases

(27) *If* $c = 0$, *then* $ax^2 + bx + c = ax^2 + bx = ax(x + \dfrac{b}{a})$ *and* $x_1, x_2 = 0, -\dfrac{b}{a}$

$Check \quad x_1, x_2 = -\dfrac{b}{2a} \pm \dfrac{\sqrt{b^2 - 4ac}}{2a} = -\dfrac{b}{2a} \pm \dfrac{\sqrt{b^2 - 0}}{2a} = -\dfrac{b}{2a} \pm \dfrac{b}{2a} = 0, -\dfrac{b}{a} \quad qed$

(28) *If* $b = 0$, *then* $ax^2 + bx + c = ax^2 + c = a(x^2 + \dfrac{c}{a})$ *and* $x_1, x_2 = \pm\sqrt{\dfrac{c}{a}}$

$Check \quad x_1, x_2 = -\dfrac{b}{2a} \pm \dfrac{\sqrt{b^2 - 4ac}}{2a} = -0 \pm \dfrac{\sqrt{0 - 4ac}}{2a} = \pm\sqrt{\dfrac{4ac}{4a^2}} = \pm\sqrt{\dfrac{c}{a}} \quad qed$

Sum and product of the roots provide an easy way to check results.

(29) $x_1 + x_2 = -\dfrac{b}{2a} + \dfrac{\sqrt{b^2 - 4ac}}{2a} - \dfrac{b}{2a} - \dfrac{\sqrt{b^2 - 4ac}}{2a} = -\dfrac{b}{a}$

(30) $x_1 x_2 = \dfrac{-b + \sqrt{b^2 - 4ac}}{2a} \times \dfrac{-b - \sqrt{b^2 - 4ac}}{2a}$

$$= \dfrac{b^2 - \left(\sqrt{b^2 - 4ac}\right)^2}{4a^2} = \dfrac{b^2 - b^2 + 4ac}{4a^2} = \dfrac{c}{a}$$

Character of the roots

roots of ax^2+bx+c are	when
(31a) *real and unequal*	$b^2-4ac>0$
(31b) *real and equal*	$b^2-4ac=0$
(31c) *complex and unequal*	$b^2-4ac<0$

3 Examples - (take care, they are different)

(32a) $x^{-6}-7x^{-3}-8=0$ → *let* $y=x^{-3}$ → $y^2-7y-8=0$

$$y_1,y_2=-\frac{b}{2a}\pm\frac{\sqrt{b^2-4ac}}{2a}=-\frac{-7}{2}\pm\frac{\sqrt{49-4(-8)}}{2}=\frac{7}{2}\pm\frac{\sqrt{81}}{2}=\frac{7}{2}\pm\frac{9}{2}=-1,8$$

(32b) $y_1=8=x^{-3}$ → $8x^3-1=0$ → $x_1=\frac{1}{2}$ *is a root by inspection*

factor $8x^3-1=(2x-1)(4x^2+2x+1)$

zeros of $4x^2+2x+1$ → $x_2,x_3=-\frac{2}{8}\pm\frac{\sqrt{4-16}}{8}=-\frac{1}{4}\pm i\frac{\sqrt{3}}{4}$

(32c) $y_2=-1=x^{-3}$ → $x^3+1=0$ → $x_1=-1$ *is a root by inspection*

factor $x^3+1=(x+1)(x^2-x+1)$

zeros of x^2-x+1 → $x_2,x_3=-\frac{-1}{2}\pm\frac{\sqrt{1-4}}{2}=\frac{1}{2}\pm i\frac{\sqrt{3}}{2}$

Solve by quadratic formula:

Problem 607 $2x^2-11x=-12$ $\qquad\qquad$ (3/2, 4)

Problem 608 $x-2x^2+15=0$ $\qquad\qquad$ (−5/2, 3)

Problem 609 $2x^2+2x=1$ $\qquad\qquad$ (1/2)(−1±√3)

Problem 610 $3x(x-1)+1=0$ $\qquad\qquad$ (1/6)(3±i√3)

Problem 611 $24w-w^2=143$ $\qquad\qquad$ (11,13)

Problem 612 $2mx=4mn-2nx+x^2$ $\qquad\qquad$ (2m, 2n)

Problem 613 $a^2x-ax^2=ab-bx$ $\qquad\qquad$ (a, b/a)

6.4 Determinant Operations

We use Cramer's rule (sidebar page 68) to solve the electric circuit node equations. The two node solution is a good test of algebraic skills.

(33)   *node 2* $i_S = y_{22}v_2 - y_{23}v_3$
      node 3 $0 = -y_{32}v_2 + y_{33}v_3$ *(general form)*

(34) $\Delta_Y = y_{22}y_{33} - y_{23}y_{32}$

Figure 508 Low Pass Filter

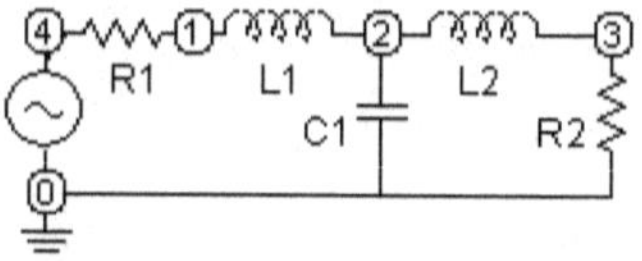

$$(35a)\quad v_2 = \frac{\begin{vmatrix} i_S & -y_{23} \\ 0 & y_{33} \end{vmatrix}}{\Delta_Y} = \frac{i_S y_{33} - (-y_{23} \times 0)}{\Delta_Y}$$

$$(35b)\quad v_3 = \frac{\begin{vmatrix} y_{22} & i_S \\ -y_{32} & 0 \end{vmatrix}}{\Delta_Y} = \frac{(y_{22} \times 0) - (-y_{32} \times i_S)}{\Delta_Y}$$

Specific form let $R = R_1 = R_2$, $L = L_1 = L_2$, $C = C_1$

$$(36a)\quad node\ 2 \quad \frac{1}{R+pL}v_S = \left(\frac{1}{R+pL} + pC + \frac{1}{pL}\right)v_2 \quad -\frac{1}{pL}v_3$$

$$(36b)\quad node\ 3 \quad\quad 0 = \quad -\frac{1}{pL}v_2 + \left(\frac{1}{pL} + \frac{1}{R}\right)v_3$$

$$(37)\quad \Delta_Y = \left(\frac{1}{R+pL} + pC + \frac{1}{pL}\right)\left(\frac{1}{pL} + \frac{1}{R}\right) - \left(-\frac{1}{pL}\right)^2 = \frac{1}{pLR}\left(2 + pCR + p^2LC\right)$$

$$(38)\quad v_2 = \frac{\begin{vmatrix} i_S & -y_{23} \\ 0 & y_{33} \end{vmatrix}}{\Delta_Y} = \frac{\frac{1}{R+pL}v_S \cdot \left(\frac{1}{pL} + \frac{1}{R}\right)}{\frac{1}{pLR}\left(2 + pCR + p^2LC\right)} = \frac{v_S}{\left(2 + pCR + p^2LC\right)}$$

$$(39)\quad v_3 = \frac{\begin{vmatrix} y_{22} & i_S \\ -y_{32} & 0 \end{vmatrix}}{\Delta_Y} = \frac{\frac{1}{R+pL}v_S \cdot \frac{1}{pL}}{\frac{1}{pLR}\left(2 + pCR + p^2LC\right)} = \frac{Rv_S}{(R+pL)\left(2 + pCR + p^2LC\right)}$$

Problem 614 Derive equations 37, 38, and 39.

Algebraic solution by determinants We show the solution to equations 40 without proof.

(40a) $a_1x + b_1y + c_1z = d_1$

(40b) $a_2x + b_2y + c_2z = d_2$

(40c) $a_3x + b_3y + c_3z = d_3$

Observe how the a, b, c columns are replaced by the d column in the solutions for x, y, z.

$$(41) \quad x = \frac{\begin{vmatrix} d_1 & b_1 & c_1 \\ d_2 & b_2 & c_2 \\ d_3 & b_3 & c_3 \end{vmatrix}}{\Delta} \qquad y = \frac{\begin{vmatrix} a_1 & d_1 & c_1 \\ a_2 & d_2 & c_2 \\ a_3 & d_3 & c_3 \end{vmatrix}}{\Delta} \qquad z = \frac{\begin{vmatrix} a_1 & b_1 & d_1 \\ a_2 & b_2 & d_2 \\ a_3 & b_3 & d_3 \end{vmatrix}}{\Delta}$$

where

$$(42) \quad \Delta = \begin{vmatrix} a_1 & b_1 & c_1 \\ a_2 & b_2 & c_2 \\ a_3 & b_3 & c_3 \end{vmatrix}$$

A fairly easy way to calculate a third order determinant, and only the third order, is to add columns 1 and 2 to the right of the determinant. The six products and their signs are indicated by the arrows.

(43) $\Delta = a_1b_2c_3 + b_1c_2a_3 + c_1a_2b_3 - c_1b_2a_3 - a_1c_2b_3 - b_1a_2c_3$

Given the equations x+2y−z=6, 2x−y+3z=−13, 3x−2y+3z=−16.

Problem 615 Solve the equations by substitution. Hint take two at a time. (x=−1, y=2, z=−3)

Problem 616 Solve the equations by addition or subtraction.

Given the equations 3p−2q+r=6, 2p+3q+2r=−1, 5q−4r=−3.

Problem 617 Solve the equations by substitution. Hint take two at a time. (p=3/2, q=−1, r=−1/2)

Problem 618 Solve the equations by addition or subtraction.

Cramer's Rule

The first subscript is the row number. The second subscript is the column number. Determinants are expanded by rows or columns.

Cramer's solutions are expansions by columns where forcing functions replace the column's elements. Note: incorporate minus signs into the a_{ij}'s.

Cramer found responses y_1, y_2 *to forcing functions* x_1, x_2

If $x_1 = a_{11}y_1 + a_{12}y_2$ *and* $x_2 = a_{21}y_1 + a_{22}y_2$

Then $\Delta = a_{11}a_{22} - a_{21}a_{12}$ *and*

$$y_1 = \frac{\begin{vmatrix} x_1 & a_{12} \\ x_2 & a_{22} \end{vmatrix}}{\begin{vmatrix} a_{11} & a_{12} \\ a_{21} & a_{22} \end{vmatrix}} = \frac{x_1 a_{22} - x_2 a_{12}}{\Delta} \qquad y_2 = \frac{\begin{vmatrix} a_{11} & x_1 \\ a_{21} & x_2 \end{vmatrix}}{\begin{vmatrix} a_{11} & a_{12} \\ a_{21} & a_{22} \end{vmatrix}} = \frac{-x_1 a_{21} + x_2 a_{11}}{\Delta}$$

And for three responses y_1, y_2, y_3 *to forcing functions* x_1, x_2, x_3

If $x_1 = a_{11}y_1 + a_{12}y_2 + a_{13}y_3$

$x_2 = a_{21}y_1 + a_{22}y_2 + a_{23}y_3$

$x_3 = a_{31}y_1 + a_{32}y_2 + a_{33}y_3$

Then $\Delta = a_{11}\Delta_{11} - a_{21}\Delta_{21} + a_{31}\Delta_{31}$ *(expansion by column 1)*

$\Delta = a_{11}(a_{22}a_{33} - a_{23}a_{32}) - a_{21}(a_{12}a_{33} - a_{13}a_{32}) + a_{31}(a_{12}a_{23} - a_{13}a_{22})$

$$y_1 = \frac{x_1\Delta_{11} - x_2\Delta_{21} + x_3\Delta_{31}}{\Delta} \qquad \textit{(expansion down column 1, rows 1, 2, 3)}$$

$$y_2 = \frac{x_1\Delta_{12} - x_2\Delta_{22} + x_3\Delta_{32}}{\Delta} \qquad \textit{(expansion down column 2, rows 1, 2, 3)}$$

$$y_3 = \frac{x_1\Delta_{13} - x_2\Delta_{23} + x_3\Delta_{33}}{\Delta} \qquad \textit{(expansion down column 3, rows 1, 2, 3)}$$

6.5 Solving Real Equations

We challenge the reader to start from **Figure 601 Low Pass Filter**
equations 45 and finish with equations
61. Do the algebraic manipulations in
equations 49 to 61 without looking at
pages 69, 70, and 71 (do problem 614).

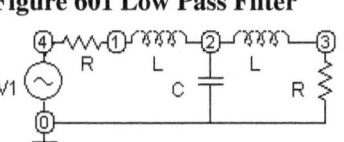

Polynomials originate in solutions to physical problems. For example
Electrical Engineers analyze the low pass filter circuit and write
equations 45a, 45b (Figure 601). Then the math takes over.

$$(45a) \quad node\ 2 \quad \frac{1}{R+pL}v_S = \left(\frac{1}{R+pL} + pC + \frac{1}{pL}\right)v_2 \quad -\frac{1}{pL}v_3$$

$$(45b) \quad node\ 3 \quad 0 = \quad -\frac{1}{pL}v_2 + \left(\frac{1}{pL} + \frac{1}{R}\right)v_3$$

The R, L, C in this, or any, electric circuit are real numbers, because they
represent physical devices. Real number coefficients result from analysis
of electric circuits, analysis of bodies in motion, and analysis of all other
physical problems. In algebra constants a, b, and c replace R, L, C.

Mathematicians start with general formats, which make a lot of sense.
Here are equations 45a, 45b in general format. Emphasis - in physical
problems the a_{jk}'s are real numbers.

$$(46a) \quad x_1 = a_{11}\,y_1 + a_{12}y_2$$

$$(46b) \quad x_2 = a_{21}y_1 + a_{22}\,y_2$$

In general simultaneous linear equations are solved by using Cramer's
rule. The not very obvious solution is

$$(47) \quad \Delta = a_{11}a_{22} - a_{21}a_{12}$$

$$(48a) \quad y_1 = \frac{x_1 a_{22} - x_2 a_{12}}{\Delta} \qquad (48b) \quad y_2 = \frac{-x_1 a_{21} + x_2 a_{11}}{\Delta}$$

We compare equations 45 and 46 to relate the x' and a's to the RLC
expressions.

$$(49) \quad \Delta = a_{11}a_{22} - a_{21}a_{12} = \left(\frac{1}{R+pL} + pC + \frac{1}{pL}\right)\left(\frac{1}{pL} + \frac{1}{R}\right) - \left(\frac{1}{pL}\right)\left(\frac{1}{pL}\right)$$

What follows is a real exercise in algebraic manipulations.

Algebra

The Arithmetic distribution law carries over to algebra. The law connects multiplication to addition.

(50) $x(y+z) = xy + xz$

Apply the distribution law once to get

(51) $\Delta = \left(\dfrac{1}{R+pL} + pC + \dfrac{1}{pL} \right)\left(\dfrac{1}{pL} \right) + \left(\dfrac{1}{R+pL} + pC + \dfrac{1}{pL} \right)\left(\dfrac{1}{R} \right) - \left(\dfrac{1}{pL} \right)\left(\dfrac{1}{pL} \right)$

Apply it again to get

(52) $\Delta = \left(\dfrac{1}{(R+pL)(pL)} + \dfrac{pC}{pL} + \dfrac{1}{p^2L^2} \right) + \left(\dfrac{1}{(R+pL)R} + \dfrac{pC}{R} + \dfrac{1}{pLR} \right) - \left(\dfrac{1}{p^2L^2} \right)$

Observe that the $1/p^2L^2$ terms cancel. Drop the parentheses.

(53) $\Delta = \dfrac{1}{(R+pL)(pL)} + \dfrac{pC}{pL} + \dfrac{1}{(R+pL)R} + \dfrac{pC}{R} + \dfrac{1}{pLR}$

Multiply *both* sides of (53) by (R+pL)(pLR), apply the distribution law, and cancel terms (the operations may not be clear at this point).

(54) $(R+pL)(pLR)\Delta = R + (R+pL)(R)(pC) + pL + (R+pL)(pL)(pC) + (R+pL)$

Add the R and pL terms

(55) $(R+pL)(pLR)\Delta = (R+pL) + (R+pL)(R)(pC) + (R+pL)(pL)(pC) + (R+pL)$

Factor out the (R+pL) term from each of the four terms. Think of this as a reverse distribution.

(56) $(R+pL)(pLR)\Delta = (R+pL)[1 + (R)(pC) + (pL)(pC) + 1]$

Divide both sides by (R+pL)(pLR)

(57) $\Delta = \dfrac{1}{pLR}[1 + (R)(pC) + (pL)(pC) + 1] = \dfrac{1}{pLR}[2 + pCR + p^2LC]$

Here is a copy of the equations

(45a) *node 2* $\dfrac{1}{R+pL}v_S = \left(\dfrac{1}{R+pL}+pC+\dfrac{1}{pL}\right)v_2 - \dfrac{1}{pL}v_3$

(45b) *node 3* $0 = -\dfrac{1}{pL}v_2 + \left(\dfrac{1}{pL}+\dfrac{1}{R}\right)v_3$

(47) $\Delta = a_{11}a_{22} - a_{21}a_{12}$

(48a) $y_1 = \dfrac{x_1 a_{22} - x_2 a_{12}}{\Delta}$ (48b) $y_2 = \dfrac{-x_1 a_{21} + x_2 a_{11}}{\Delta}$

Form the y_1 numerator.

(58) $x_1 a_{22} - x_2 a_{12} = \dfrac{1}{R+pL}v_S\left(\dfrac{1}{pL}+\dfrac{1}{R}\right) - 0\cdot\left(-\dfrac{1}{pL}\right)$

Add the 2 fractions the same way one would in Arithmetic.

(59) $\left(\dfrac{1}{pL}+\dfrac{1}{R}\right) = \dfrac{1}{pL}\dfrac{R}{R} + \dfrac{1}{R}\dfrac{pL}{pL} = \dfrac{R+pL}{pLR}$

Substitute 59 in equation 58, and form the y_1, y_2 numerators.

(60a) $x_1 a_{22} - x_2 a_{12} = \dfrac{1}{R+pL}v_S\dfrac{R+pL}{pLR} - 0\cdot\left(-\dfrac{1}{pL}\right) = \dfrac{v_S}{pLR}$

(60b) $-x_1 a_{21} + x_2 a_{11} = -\dfrac{1}{R+pL}v_S\left(-\dfrac{1}{pL}\right) - 0 = \dfrac{v_S}{(R+pL)pL}$

Apply Cramer's Rule

(61a) $v_2 = \dfrac{x_1 a_{22} - x_2 a_{12}}{\Delta_Y} = \dfrac{\dfrac{1}{pLR}v_S}{\dfrac{1}{pLR}\left(2+pCR+p^2LC\right)} = \dfrac{v_S}{\left(2+pCR+p^2LC\right)}$

(61b) $v_3 = \dfrac{-x_1 a_{21} + x_2 a_{11}}{\Delta_Y} = \dfrac{\dfrac{1}{(R+pL)pL}v_S}{\dfrac{1}{pLR}\left(2+pCR+p^2LC\right)}$

$\dfrac{v_3}{v_S} = \dfrac{R}{(R+pL)\left(2+pCR+p^2LC\right)}$

Algebra

6.6 Elimination Operations

This non-trivial example shows the significant role of elimination by addition or subtraction, and by substitution in solving equations. EE circuit analysis produces equations 62a to 62f. The v's and i's are variables, the R's and g_m are constants. Algebraic analysis solves for the signal transmission $T = v_3/v_1$ from input to output.

Again we challenge the reader to start from equation 62a and finish with equation 62g.

Figure 608 V to I Circuit

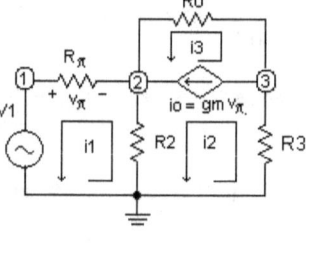

(62a) $\quad v_1 = i_1 R_\pi + (i_1 - i_2)R_2 \qquad mesh\ 1$

(62b) $\quad 0 = (i_2 - i_1)R_2 + v_x + i_2 R_3 \quad mesh\ 2$

(62c) $\quad 0 = -v_x + i_3 R_0 \qquad\qquad mesh\ 3$

(62d) $\quad i_0 = g_m v_\pi = i_2 - i_3$

(62e) $\quad v_\pi = i_1 R_\pi$

(62f) $\quad v_3 = i_2 R_3 \quad\rightarrow\quad T = \dfrac{v_3}{v_1} = \dfrac{i_2 R_3}{v_1} = R_3 \dfrac{i_2}{v_1}$

We need an expression for i_2/v_1 (62f). Voltage v_1 in equation 62a is a function of i_1 and i_2. We need to eliminate i_1 so that v_1 is only a function of i_2. Then we have a solution for transmission T (62f).

We do not need v_x so add 62c to 62b to eliminate the v_x *terms.*

(62b) $\qquad (i_2 - i_1)R_2 + v_x + i_2 R_3 \qquad\qquad mesh\ 2$

(62c) $\qquad\qquad\qquad -v_x \qquad + i_3 R_0 \quad mesh\ 3$

(62b+c) $\quad (i_2 - i_1)R_2 + 0\ \ + i_2 R_3 + i_3 R_0 \quad mesh\ 2+3$

Replace 62b and 62c with 62b+c

(62a) $\qquad v_1 = i_1 R_\pi + (i_1 - i_2)R_2 \qquad\qquad mesh\ 1$

(62b+c) $\quad 0 = (i_2 - i_1)R_2 + i_2 R_3 + i_3 R_0 \quad mesh\ 2+3$

(62d) $\quad g_m v_\pi = i_0 = i_2 - i_3$

Use the distribution law to collect terms in 62a and 62b+c so that currents are *factors* as in $i_1 R_\pi + i_1 R_2 = i_1(R_\pi + R_2)$

(64a) $\quad v_1 = i_1(R_\pi + R_2) - i_2 R_2 \qquad\qquad mesh\ 1$

(64b) $\quad 0 = -i_1 R_2 + i_2(R_2 + R_3) + i_3 R_0 \quad mesh\ 2+3$

(62d) $\quad g_m v_\pi = i_0 = i_2 - i_3$

72

(62d) $\quad g_m v_\pi = i_2 - i_3$

(65a) $\quad$ *add* i_3 *to both sides* $\quad i_3 + g_m v_\pi = i_2 - i_3 + i_3 \quad \rightarrow \quad i_3 + g_m v_\pi = i_2$

(65b) $\quad$ *sub* $g_m v_\pi$ *from both sides* $\quad i_3 + g_m v_\pi - g_m v_\pi = i_2 - g_m v_\pi \quad \rightarrow \quad i_3 = i_2 - g_m v_\pi$

(65c) $\quad$ *sub* $i_1 R_\pi$ *for* $v_\pi \quad \rightarrow \quad i_3 = i_2 - g_m i_1 R_\pi$

Now substitute the i_3 equation 65c for i_3 in 64b.

(66a) $\quad v_1 = i_1(R_\pi + R_2) - i_2 R_2 \qquad$ *mesh 1*

(66b) $\quad 0 = -i_1 R_2 + i_2(R_2 + R_3) + (i_2 - g_m R_\pi i_1)R_0 \quad$ *mesh 2+3*

Collect terms again.

(67a) $\quad v_1 = i_1(R_\pi + R_2) - i_2 R_2 \qquad\qquad\qquad$ *mesh 1*

(67b) $\quad 0 = -i_1(R_2 + g_m R_\pi R_0) + i_2(R_0 + R_2 + R_3) \quad$ *mesh 2+3*

Equation 67a shows v_1 depends on i_1 and i_2, and equation 67b relates i_1 to i_2. In equation 62f we need i_2/v_1 so eliminate i_1 by substitution.

(67b) $\quad 0 = -i_1(R_2 + g_m R_\pi R_0) + i_2(R_0 + R_2 + R_3) \quad$ *mesh 2+3*

(67c) $\quad$ *move* i_1 *term to left side of 04b* $\quad i_1(R_2 + g_m R_\pi R_0) = i_2(R_0 + R_2 + R_3)$

(67d) $\quad$ *divide both sides by* $(R_2 + g_m R_\pi R_0) \quad \rightarrow \quad i_1 = i_2 \dfrac{(R_0 + R_2 + R_3)}{(R_2 + g_m R_\pi R_0)}$

Substitute 67d into 67a to eliminate i_1.

(68) $\quad v_1 = i_2 \dfrac{(R_0 + R_2 + R_3)}{(R_2 + g_m R_\pi R_0)}(R_\pi + R_2) - i_2 R_2$

Substitute the reciprocal of v_1/i_2 into 62f.

(62f) $\quad \dfrac{v_3}{v_1} = \dfrac{i_2}{v_1} R_3 = \dfrac{1}{\dfrac{(R_0 + R_2 + R_3)}{(R_2 + g_m R_\pi R_0)}(R_\pi + R_2) - R_2} R_3$

The form of this result can be improved by multiplying numerator and denominator by $(R_2 + g_m R_\pi R_0)$ and multiplying out the denominator.

(62g) $\quad \dfrac{v_3}{v_1} = \dfrac{i_2}{v_1} R_3 = \dfrac{(R_2 + g_m R_\pi R_0)R_3}{(R_0 + R_2 + R_3)(R_\pi + R_2) - (R_2 + g_m R_\pi R_0)R_2}$

7 Exponents

An *exponent n* is a symbol written above, and on the right of, another symbol known as the *base x* as in x^n. Another exponent form is b^x with base b and variable x as the exponent (Chapter 9).

The expression x^n is referred to as an *algebraic power function*; specifically the nth power of x. All arithmetic operations apply to exponents. *The base and exponent can be any type of numbers*. However we restrict exponents to real numbers.

7.1 Positive Integer Exponents

The symbol x^n is read as "x to the nth power" or "x to the nth", or whatever you choose to say (x^2 is x squared, x^3 is x cubed). When n is a positive integer the world says *exponent n is the nth power of base x*. And, when n is an integer, *the symbol x^n represents the product of n factors each equal to x.*

(1) $x^n = x \cdot x \cdot x \cdots x \cdot x, \quad n \ \ factors \ x$

Laws of Positive Integer Exponents The laws of Positive Integer exponents extend algebraic processes to include *powers* of variables.

Addition of powers

(2) $x^n \cdot x^m = x^{n+m}$

Proof $x^n \cdot x^m = (x \cdot x \cdots x, \ n \ \ factors \ x)(x \cdot x \cdots x, \ m \ \ factors \ x)$

$\qquad = x \cdot x \cdots x, \ n+m \ \ factors \ x$

$\qquad = x^{n+m}$

Example $3^5 7^2 3^3 = 3^8 7^2$ *exponents add only when bases are equal*

Power of a power of x

(3) $(x^m)^n = x^{mn}$

Proof $(x^m)^n = x^m \cdot x^m \cdots x^m, n \ \ factors \ x^m$

$\qquad = x^{m+m+\cdots+m \ n \ terms \ m} = x^{mn}$

Example $(4^3)^8 = 4^{24}$ *this is a product of eight 4^3 where eight 3's add*

Power of a product of powers of x and y

(4) $(xy)^n = x^n y^n$

Proof $(xy)^n = xy \cdot xy \cdots xy,\ n\ factors\ xy$

$$= (x \cdot x \cdots x,\ n\ factors\ x)(y \cdot y \cdots y,\ n\ factors\ y)$$

$$= x^n y^n$$

Example $(2 \times 9)^5 = 2^5 9^5$

product of five 2×9's is product of five 2's & five 9's

Power of a fraction of powers of x and y

(5) $\left(\dfrac{x}{y}\right)^n = \dfrac{x^n}{y^n}$

Proof $\left(\dfrac{x}{y}\right)^n = \left(\dfrac{x}{y}\right) \cdot \left(\dfrac{x}{y}\right) \cdots \left(\dfrac{x}{y}\right),\ n\ factors\ \left(\dfrac{x}{y}\right)$

$$= \dfrac{x \cdot x \cdots x,\ n\ factors\ x}{y \cdot y \cdots y,\ n\ factors\ y}$$

$$= \dfrac{x^n}{y^n}$$

Example $\left(\dfrac{5}{6}\right)^8 = \left(\dfrac{5}{6}\right)\ multiplied\ 8\ times = \dfrac{5\ multiplied\ 8\ times}{6\ multiplied\ 8\ times} = \dfrac{5^8}{6^8}$

Power of a ratio of powers of x where m>n

(6) $\dfrac{x^m}{x^n} = x^{m-n}\quad where\ m > n,\ \ x \neq 0$

Proof $\dfrac{x^m}{x^n} = \dfrac{x \cdot x \cdots x,\ m\ factors\ x}{x \cdot x \cdots x,\ n\ factors\ x}$

$$= \dfrac{(x \cdot x \cdots x,\ m-n\ factors\ x)(x \cdot x \cdots x,\ n\ factors\ x)}{(x \cdot x \cdots x,\ n\ factors\ x)}$$

$$= x \cdot x \cdots x,\ m-n\ factors\ x = x^{m-n}$$

Example $\dfrac{4^7}{4^5} = \dfrac{4^7}{4^5} \dfrac{4^{-5}}{4^{-5}} = \dfrac{4^7 \times 4^{-5}}{4^5 \times 4^{-5}} = \dfrac{4^2}{4^0} = 4^2$

Algebra

Power of a ratio of powers of x where n>m

(7) $\quad \dfrac{x^m}{x^n} = \dfrac{1}{x^{n-m}} \quad$ where $n > m, \ x \neq 0$

$Proof \quad \dfrac{x^m}{x^n} = \dfrac{x \cdot x \cdots x, \ m \ \ factors \ x}{x \cdot x \cdots x, \ n \ \ factors \ x}$

$$= \dfrac{(x \cdot x \cdots x, \ m \ \ factors \ x)}{(x \cdot x \cdots x, \ n-m \ \ factors \ x)(x \cdot x \cdots x, \ m \ \ factors \ x)}$$

$$= \dfrac{1}{x \cdot x \cdots x, \ n-m \ \ factors \ x} = \dfrac{1}{x^{n-m}}$$

$Example \quad \dfrac{4^3}{4^7} = \dfrac{4^3}{4^7} \dfrac{4^{-3}}{4^{-3}} = \dfrac{4^3 \times 4^{-3}}{4^7 \times 4^{-3}} = \dfrac{4^0}{4^4} = \dfrac{1}{4^4}$

Examples of Positive Integer Exponents

(8) $\quad x^4 \cdot x^{13} = x^{17} \qquad (x^5)^7 = x^{35} \qquad (xy)^{12} = x^{12} y^{12}$

$\qquad \left(\dfrac{x}{y}\right)^8 = \dfrac{x^8}{y^8} \qquad\qquad \dfrac{x^7}{x^5} = x^2 \qquad\qquad \dfrac{x^3}{x^{13}} = \dfrac{1}{x^{10}}$

7.2 Fractional Exponents (Roots)

Let m=1/q where q is any positive integer. If 1/q, as an exponent, is to obey the addition law, then

(9) $\quad \left(x^{\frac{1}{q}}\right)^q = x^{\frac{1}{q}} \cdot x^{\frac{1}{q}} \cdots x^{\frac{1}{q}}, \ q \ \ factors \ x^{\frac{1}{q}} = x^{\overset{q \ terms \ \frac{1}{q}}{\frac{1}{q}+\frac{1}{q}+\cdots+\frac{1}{q}}} = x^{\frac{q}{q}} = x^1 = x$

$\qquad \therefore \ x^{\frac{1}{q}} \ is \ defined \ as \ the \ qth \ root \ of \ x$

$Example \quad \left(x^{\frac{1}{6}}\right)^6 = product \ of \ six \left(x^{\frac{1}{6}}\right) = x \ because \ sum \ of \ six \ \frac{1}{6} \ equals \ 1$

Changing from q to p factors we get

(10) $\quad \left(x^{\frac{1}{q}}\right)^p = x^{\frac{1}{q}} \cdot x^{\frac{1}{q}} \cdots x^{\frac{1}{q}}, \ p \ \ factors \ x^{\frac{1}{q}} = x^{\overset{p \ terms \ \frac{1}{q}}{\frac{1}{q}+\frac{1}{q}+\cdots+\frac{1}{q}}} = x^{\frac{p}{q}}$

$Example \quad \left(x^{\frac{1}{6}}\right)^5 = product \ of \ five \left(x^{\frac{1}{6}}\right) = x^{\frac{5}{6}}$

$because \ sum \ of \ five \ \frac{1}{6} \ equals \ \frac{5}{6}$

Changing the exponent to p/q we get

(11) $\left(x^{\frac{p}{q}}\right)^{q} = x^{\frac{p}{q}} \cdot x^{\frac{p}{q}} \cdots x^{\frac{p}{q}}, q\ \textit{factors}\ x^{\frac{p}{q}} = x^{\frac{p}{q}+\frac{p}{q}+\cdots+\frac{p}{q}}\ \textit{q terms}\ \frac{p}{q} = x^{q\frac{p}{q}} = x^{p}$

Example $\left(x^{\frac{2}{7}}\right)^{7} = $ *product of seven* $\left(x^{\frac{2}{7}}\right) = x^{2}$

because sum of seven $\frac{2}{7}$ *equals 2*

We use fractional exponents to avoid radicals such as $^{3}\sqrt{2^{5}} = (2^{5})^{1/3} = 2^{5/3}$

We avoid radicals like the plague, because radicals are an easy source of errors whereas exponents are straightforward.

7.3 Exponent Zero

If zero, as an exponent, is to obey the addition law, then when $x \neq 0$

(12) $\quad x^{n}x^{0} = x^{n+0} = x^{n} \quad \Rightarrow \quad 1 = \dfrac{x^{n}}{x^{n}} = x^{n-n} = x^{0} \quad \Rightarrow \quad \underline{x^{0} = 1}$

Example   since $\dfrac{a}{a} = 1$

$\dfrac{10^{3}}{10^{3}} = \dfrac{10^{3}}{10^{3}} \times \dfrac{10^{-3}}{10^{-3}} = \dfrac{10^{3-3}}{10^{0}} = 10^{3-3} = 10^{0} = 1$

7.4 Negative Exponents

(13) $\quad x^{n}x^{-n} = x^{n-n} = x^{0} = 1 \quad \Rightarrow \quad x^{-n} = \dfrac{1}{x^{n}}$

$\quad\quad\textit{Example}\quad \dfrac{1}{x^{5}} = \dfrac{1}{x^{5}} \cdot \dfrac{x^{-5}}{x^{-5}} = \dfrac{x^{-5}}{x^{0}} = x^{-5}$

Algebra

Problems 7

Simplify

1. $2^8 4^5$ 2. $27^5 / 3^{11}$ 3. $25^{x+2} / 5^{x-1}$ 4. $9^{2m}(3^m)^{m+1}$ 5. $4^2 2^{3n} / 8^{n+2}$

6. $\dfrac{c^{x^2}}{c^{x^2(x+1)}}$ 7. $\dfrac{(a^{2x-y})^{x+2y}}{(a^{2x+y})^{x-2y}}$ 8. $\dfrac{x^{(a^2-9)}}{x^{a-3}}$ 9. $\dfrac{a^{m-2n}a^{3(m+n)}}{a^{2m-n}}$ 10. $\left(\dfrac{b^{2x-3}}{b^{2x+3}}\right)^{x+1}$

Find the values. Hint write negative numbers −n as −1×n.

1. $81^{\frac{1}{2}}$ 2. 81^0 3. $0^{\frac{1}{2}}$ 4. $64^{\frac{1}{4}}$ 5. $27^{\frac{1}{3}}$ 6. $27^{\frac{2}{3}}$

7. $27^{\frac{4}{3}}$ 8. $16^{\frac{1}{4}}$ 9. $16^{\frac{3}{4}}$ 10. $\left(\frac{9}{25}\right)^{\frac{1}{2}}$ 11. $\left(\frac{9}{25}\right)^{\frac{3}{2}}$ 12. $0.04^{\frac{1}{2}}$

13. $0.216^{\frac{2}{3}}$ 14. $(-8)^{\frac{1}{3}}$ 15. $\left(-\frac{1}{32}\right)^{\frac{2}{5}}$ 16. $(-4)^3$ 17. 7^{-2} 18. $\left(\frac{2}{3}\right)^{-3}$

Convert to positive exponents.

1. x^{-2} 2. $x^{\frac{3}{4}}x^{-\frac{1}{2}}$ 3. $(x^{-\frac{1}{2}})^{-\frac{5}{3}}$ 4. $(x^{-\frac{1}{2}})^{-\frac{5}{3}}$ 5. $(-x^{-\frac{5}{6}})^{-\frac{1}{5}}$

6. $2x^{-1}y^{-2}$ 7. $\dfrac{3x^{-3}}{yz^{-4}}$ 8. $\dfrac{2x^{-1}y^4}{3^{-2}x^3y^{-5}}$ 9. $\dfrac{2^{-1}b^3c^{-\frac{2}{3}}}{5b^{-\frac{1}{4}}c^2}$ 10. $\dfrac{3x^{-\frac{2}{5}}y^{-\frac{3}{2}}}{2^{-2}x^{-\frac{1}{2}}y^{-\frac{5}{6}}}$

Convert denominator to 1.

1. $\dfrac{3x^2}{z^{-3}}$ 2. $\dfrac{3a}{x^4z^{-3}}$ 3. $\dfrac{x^2}{4y^{-\frac{2}{3}}}$ 4. $\dfrac{x^{(a^2-9)}}{x^{a-3}}$ 5. $\dfrac{c^{x^2}}{c^{x^2(x+1)}}$

Simplify

1. $x^{-1}-y^{-1}$ 2. $(x^{-3}+y^{-3})^{-1}$ 3. $\dfrac{1}{x^{-2}-y^{-2}}$

4. $\dfrac{x^{-1}+y^{-1}}{y^{-2}-x^{-2}}$ 5. $\dfrac{xy^{-2}+x^{-2}y}{x^{-1}-y^{-1}}$ 6. $\left(\dfrac{xy^{-1}-x^{-1}y}{xy^{-2}-x^{-2}y}\right)^{-1}$

7. $(x^{\frac{2}{3}}-y^{\frac{2}{3}})^3$ 8. $\dfrac{(a^2+x^2)^{-\frac{3}{5}}-x^{-2}}{(a^2+x^2)^{\frac{3}{5}}-x^2}$ 9. $\dfrac{3+9x(9x^2+1)^{-\frac{1}{2}}}{3x+(9x^2+1)^{\frac{1}{2}}}$

10. $\dfrac{(1-4x^2)^{\frac{1}{2}}-2x}{4x(1-4x^2)^{-\frac{1}{2}}-2}$ 11. $\dfrac{\frac{1}{5}x^{-2}-x^{-1}}{(x\div\frac{1}{125}x^{-2})^{-\frac{2}{3}}-1}$ 12. $\dfrac{x^2(x^2-a^2)^{-\frac{1}{2}}-(x^2-a^2)^{-\frac{1}{2}}}{(x^2-a^2)}$

78

8 The Binomial Theorem for any Index

The Binomial Theorem shows how to expand $(x+a)^n$ when n is an integer thereby avoiding the tedious process of multiplying by $(x+a)$ n−1 times.

Furthermore, the Binomial Theorem shows how to expand $(x+a)^n$ when n is any number, positive, negative, integral or fractional where x and a can be any numbers. The Binomial Theorem has many applications such as calculating $(1.08)^4$ when formed as $(1+0.08)^4$ to 5 significant figures.

8.1 Product of Factors (n integer)

We know how to use the distributive law to multiply factors such as

(1) $(x+a)(x+b) = x(x+b)+a(x+b) = x^2 + xb + ax + ab$

When we multiply by a third factor we get

(2) $(x+a)(x+b)(x+c) = x^3 + (a+b+c)x^2 + (ab+bc+ca)x + abc$

Actual multiplication is avoided if we proceed another way.
(1) Take x from 3 factors. The product is x^3.
(2) Take x from 2 factors and the constant from the third factor in 3 ways to get ax^2, bx^2, cx^2.
(3) Take x from 1 factor and the constant from 2 factors in 3 ways to get abx, bcx, cax.
(4) Take the constant from 3 factors in 1 way to get abc.
The sum of these terms confirms (2).

A particular case of (2) is produced by letting $a=b=c$ to get

(3) $(x+a)^3 = x^3 + 3x^2a + 3xa^2 + a^3$

In the same way we can show that

(4) $(x+a)^4 = x^4 + 4x^3a + 6x^2a^2 + 4xa^3 + a^4$

Avoiding multiplying out is an easier way to expand $(x+a)^n$ when n is an integer by asking the question "how many ways to choose x from a group of factors?" This brings us to permutations and combinations.

Algebra

8.2 Permutations

On many occasions we need to know the number of different ways events can occur. Events such as dealing a poker hand, selecting of 2 out of 7 items, and our immediate concern, how many ways to choose x and a from a group of factors. The two basic kinds of the number of different ways to do these operations are permutations and combinations.

Three letters abc taken 3 at a time are written as $abc, acb, bac, bca, cab, cba$, each of which is referred to as an *order*. Thus the 3 letters abc can be written down in 6 different orders. Chose 1 of 3 letters for the first position, 1 of 2 for the second position, and 1 of 1 for the third position. I.e. $abc, acb, bac, bca, cab, cba$. *The sequence of letters defines the order.* Clearly there are $3 \times 2 \times 1$ ways to fill the positions. The number $3 \times 2 \times 1$ is referred to as 3! (3 factorial), $4 \times 3 \times 2 \times 1$ is referred to as 4!, and so forth.

Definitions   *If n is a positive integer, then $n!$ (n fractorial) is defined as*
$$n! = n(n-1)(n-2)(n-3)\cdots 3 \cdot 2 \cdot 1 \quad and \quad 0! = 1$$
Example $5! = 5 \cdot 4 \cdot 3 \cdot 2 \cdot 1 = 120$

Theorem 1   *Given k positions, numbered $1, 2, \ldots, k$, and n letters ($n \geq k$),*
there are $n(n-1)(n-2)\ldots(n-\{k-1\})$
different ways of assigning k of the n letters to the k positions

Example   *If $n = 4$, $k = 3$, then $n-(k-1) = 4-2 = 2$ and $(4)(3)(2) = 24$*

The number of different ways is the number of permutations $_nP_k$
of n items taken k at a time

(5) $_nP_k = n(n-1)(n-2)\ldots(n-\{k-1\}) \times \dfrac{(n-k)!}{(n-k)!} = \dfrac{n!}{(n-k)!}$

Example $_6P_2 = 6(6-\{2-1\}) = 6(6-1) = \dfrac{6!}{(6-2)!} = \dfrac{6 \cdot 5 \cdot 4 \cdot 3 \cdot 2 \cdot 1}{4 \cdot 3 \cdot 2 \cdot 1} = 6 \cdot 5 = 30$

The example shows that the factorial form is not necessary. However the factorial form provides a compact formula for $_nP_k$.

8.3 Combinations

Three letters *abc* taken 2 at a time are written *without regard to order* as *ab, ac, bc*. The different ways are referred to as *combinations*.

Three letters *abc* taken 3 at a time can be written down *without regard to order* in 1 way *abc*.

Theorem 2 Given n letters, the number of different ways to select k of the n letters $(n \geq k)$ with no regard to order is

$$_nC_k = \frac{n(n-1)(n-2)....(n-\{k-1\})}{k!} = \frac{_nP_k}{k!}$$

Proof Let $_nC_k$ be the number of different ways to select without regard to order. Any one selection of k letters can be arranged in $k!$ different orders. Therefore there are $k! \times {_nC_k}$ ways to fill k positions. By Theorem 1 the number of different ways to select *with* regard to order is $_nP_k$. Then we have $k! \times {_nC_k} = {_nP_k}$ so that $_nC_k = {_nP_k} / k!$.

Definition

The number of ways in which k letters can be selected from n letters $(n \geq k)$, with no regard to order, is referred to as the number of combinations $_nC_k$ of n letters taken k at a time

The number of combinations

$$(6) \quad _nC_k = \frac{n(n-1)(n-2)....(n-\{k-1\})}{k!} \times \frac{(n-k)!}{(n-k)!} = \frac{n!}{k!(n-k)!}$$

Example $\quad _6C_2 = \dfrac{6!}{2!(6-2)!} = \dfrac{6 \cdot 5 \cdot 4 \cdot 3 \cdot 2 \cdot 1}{2 \cdot 1 \cdot 4 \cdot 3 \cdot 2 \cdot 1} = \dfrac{6 \cdot 5}{2} = 15$

There are a multitude of relations involving the $_nC_k$. One such relation applicable to the Binomial Theorem is

$$(7) \quad _nC_k = \frac{n!}{k!(n-k)!} = \frac{n!}{(n-k)!k!} = {_nC_{n-k}}$$

Algebra

8.4 The Binomial Theorem (integer index)

Binomial Theorem (n integer) When n is a positive integer

$$(x+a)^n = x^n + {}_nC_1 x^{n-1}a^1 + ... + {}_nC_k x^{n-k}a^k + ... + {}_nC_{n-1}x^1 a^{n-1} + a^n$$

where ${}_nC_k = \dfrac{n!}{k!(n-k)!}$ *so that*

$$(x+a)^n \equiv x^n + nx^{n-1}a^1 + \frac{n(n-1)}{2!}x^{n-2}a^2 + ... + \frac{n!}{k!(n-k)!}x^{n-k}a^k + ... + nx^1 a^{n-1} + a^n$$

Proof Consider the product $(x+a)(x+a)(x+a)...(x+a)$ *n factors*

The product is the sum of all the products we can form by selecting one term from each parens and multiplying.

(1) Take x from each of n parens, and multiply. The term is x^n.

(2) Take a from one paren, and x from each of n–1 parens, and multiply. the term is $x^{n-1}a$.

(3) With $k<n$ we select k out of n parens in ${}_nC_k$ ways. From the k parens we take a out of each of them, and we take x out of the remaining $n-k$ parens. The product is $x^{n-k}a^k$. Therefore the term is ${}_nC_k x^{n-k}a^k$.

(4) Take a from each of n parens, and multiply. The term is a^n.

Thus the sum of all the terms we get by multiplying out the parens is

$$x^n + {}_nC_1 x^{n-1}a^1 + ... + {}_nC_k x^{n-k}a^k + ... + {}_nC_{n-1}x^1 a^{n-1} + a^n$$

The ${}_nC_k$ are referred to as the binomial coefficients. The name binomial is a carry over from the technical term binomial for $x+y$.

Expand by the binomial theorem (calculating the ${}_nC_k$.)

Problem 801 $(a+b)^4$

Problem 802 $\left(\frac{1}{2}b - 3x^3\right)^3$

Problem 803 $(e^x + e^{-x})^9$

8.5 Factorial n for any Index

Factorial n terminates only when n is a positive integer. Factorial n does *not* terminate when n is a negative integer or a fraction.

Definition If n is a positive integer, then n! is defined as

$$n! = n(n-1)(n-2)(n-3)...3 \cdot 2 \cdot 1$$

and $0! = 1$, *because* $n! = n(n-1)!$ *so that* $1! = 1(1-1)! = 0!$

(8) *If n is a positive integer, then n! terminates. The last number is 1*

$$7! = 7(7-1)(7-2)(7-3)(7-4)(7-5)(7-6) = 7 \times 6 \times 5 \times 4 \times 3 \times 2 \times 1$$

(9) *If n is a negative integer, then n! does not terminate*

$$-3! = -3(-3-1)(-3-2)(-3-3).... = -3 \times -4 \times -5 \times -6 \times ....$$

(10) *If n is a fraction, then n! does not terminate*

$$\frac{1}{2}! = \frac{1}{2}\left(\frac{1}{2}-1\right)\left(\frac{1}{2}-2\right)\left(\frac{1}{2}-3\right)\cdots = \frac{1}{2} \times -\frac{1}{2} \times -\frac{3}{2} \times -\frac{5}{2} \times \cdots$$

$$-\frac{1}{2}! = -\frac{1}{2}\left(-\frac{1}{2}-1\right)\left(-\frac{1}{2}-2\right)\left(-\frac{1}{2}-3\right)\cdots = -\frac{1}{2} \times -\frac{3}{2} \times -\frac{5}{2} \times -\frac{7}{2} \times \cdots$$

8.6 Introduction of Infinite Series

Infinite series are introduced so that we can learn how to deal with an n! that does not terminate. When we divide $1-x^3$ by $1-x$ the long division terminates, because the x^3 term drops down..

(11)

$$
\require{enclose}
\begin{array}{r}
1+x+x^2 \\
1-x \enclose{longdiv}{1 -x^3} \\
\underline{1-x } \\
x -x^3 \\
\underline{x-x^2 } \\
x^2 - x^3 \\
\underline{x^2 - x^3} \\
0
\end{array}
$$

When we divide 1 by 1−x the long division process does not terminate, because there is no x^k term to drop down. The quotient appears as

(12a) $1+x+x^2+x^3+....$ *ad infinitum*

(12b)

$$
\begin{array}{r}
1+x+x^2+x^3+\cdots \\
1-x\overline{)1} \\
\underline{1-x} \\
x \\
\underline{x-x^2} \\
x^2 \\
\underline{x^2-x^3} \\
x^3 \\
\underline{x^3-x^4} \\
x^4
\end{array}
$$

How do we interpret this *series* of terms so that the result makes sense? We know we can add a finite number of terms, so we do that.

(13a) $S(n)=1+x+x^2+x^3+....+x^{n-1}$

(13b) $S(n)-xS(n)=1-x^n$

(13c) $(1-x)S(n)=1-x^n \implies S(n)=\dfrac{1-x^n}{1-x}=\dfrac{1}{1-x}-\dfrac{x^n}{1-x}$

Convergent series A series of terms is convergent if the sum of the terms is finite as n increases to infinity (infinity is an alias for 'as large as we please'). An alternative to saying "the sum of the terms is finite as n increases to infinity" is the *limit* symbolism.

(14) *If* $x<1$ *and* $x>-1$, *then the magnitude of* x *is less than 1 and* $\lim\limits_{n\to\infty}x^n=0$

Taking the sum S(n) to the limit we get

(15) $\lim\limits_{n\to\infty}S(n)=\lim\limits_{n\to\infty}\left(\dfrac{1}{1-x}-\dfrac{x^n}{1-x}\right)=\dfrac{1}{1-x}$

In this sense we can use the infinite series to represent 1/(1−x).

(16) $\dfrac{1}{1-x}=1+x+x^2+x^3+x^4+....$ $\rightarrow$ $|x|<1$

The series *converges* to 1/(1−x). We refer to the series as a *convergent series*. When the magnitude of x is greater than 1 the series is said to *diverge*. When x=1 the series does not exist.

8.7 The Binomial Theorem (any index)

Binomial Theorem

Let n be any number, positive or negative, integral or fractional.

Then the function $(1+x)^n$ *is represented by the binomal series*

$$1+nx+\frac{n(n-1)}{2!}x^2+\frac{n(n-1)(n-2)}{3!}x^3+....+\frac{n!}{k!(n-k)!}x^k+....$$

Notes
(1) When n is a positive integer the series terminates and the sum is $(1+x)^n$ for all x.

(2) When n is not a positive integer the series is an infinite series that represents $(1+x)^n$ when $-1<x<1$.

(3) Proof of the binomial theorem is outside the scope of this text, because knowledge of the general theory of infinite series is required. The theorem may be used with confidence provided however that the user remembers that the sum of *k* terms of the series *approaches* the value of $(1+x)^n$ as *k* increases. In other words we do not try to add up an infinite number of terms.

Examples

Let $n=\frac{1}{2}$

$$(1+x)^{\frac{1}{2}}=1+\frac{1}{2}x+\frac{\frac{1}{2}\left(\frac{1}{2}-1\right)}{2\cdot1}x^2+\frac{\frac{1}{2}\left(\frac{1}{2}-1\right)\left(\frac{1}{2}-2\right)}{3\cdot2\cdot1}x^3+....$$

$$(1+x)^{\frac{1}{2}}=1+\frac{1}{2}x-\frac{1}{8}x^2+\frac{1}{16}x^3+....$$

Let $n=-1$ *and* $x\to-x$

$$(1-x)^{-1}=1+(-1)(-x)+\frac{-1(-1-1)}{2\cdot1}(-x)^2+\frac{-1(-1-1)(-1-2)}{3\cdot2\cdot1}(-x)^3+....$$

$$\frac{1}{1-x}=1+x+x^2+x^3+....$$

Algebra

Problems 8

804 Simplify

1. $\dfrac{5!}{3!}$ 2. $\dfrac{9!}{6!}$ 3. $\dfrac{6!\cdot 8!}{7!\cdot 9!}$ 4. $\dfrac{4!+5!}{3!\cdot 4!}$

5. $\dfrac{5!\cdot 6!}{9!-7!}$ 6. $\dfrac{(n-1)!}{n!}$ 7. $\dfrac{p!}{(p-2)!}$ 8. $\dfrac{2k!}{(2k)!}$

9. $\dfrac{n!}{(n-r)!}$ 10. $\dfrac{(n-k-1)!}{(n-k+1)!}$ 11. $\dfrac{(n+1)!-n!}{n!+(n-1)!}$ 12. $\dfrac{[(2n+1)!]^2}{(2n)!(2n+2)!}$

13. *Show that* n!, n > 1, *is always an even number.*

14. *Show that* $\dfrac{n(n-1)(n-2)\cdots(n-r+1)}{r!} = \dfrac{n!}{r!(n-r)!}$

15. *Show that* $\dfrac{n!}{k!(n-k)!} + \dfrac{n!}{(k+1)!(n-k-1)!} = \dfrac{(n+1)!}{(k+1)!(n-k)!}$

805 Expand by the binomial theorem and simplify each term

1. $(x+y)^4$ 2. $(x-2y)^6$ 3. $(\tfrac{1}{2}x-3y^3)^3$ 4. $(x^{\frac{1}{2}}+y^{\frac{1}{2}})^5$

806 Find the first four terms of the binomial expansion and simplify each term.

1. $(x^{\frac{2}{3}}-\tfrac{1}{3}y^{-2})^{11}$ 2. $(x^{-3}+\tfrac{2}{3}y^{\frac{3}{2}})^{10}$ 3. $(x^4-2y^{-4})^{\frac{1}{4}}$ 4. $(8x^3+3y^2)^{\frac{2}{3}}$

807 Find the last four terms of the binomial expansion and simplify each term.

1. $(\tfrac{1}{3}x^2+y^{-2})^{10}$ 2. $(\tfrac{1}{2}x^{-1}-ay^{\frac{1}{2}})^{12}$ 3. $(\tfrac{2}{5}x^{\frac{1}{3}}+y^{\frac{3}{2}})^{11}$

808 Find and simplify the specified term

1. *5th term of* $(e^{2x}+e^{-2x})^{10}$

2. *5th term of* $(x^3+3y^3)^{\frac{4}{3}}$

3. x^7 *term of* $(\tfrac{1}{2}+x)^{13}$

4. $x^{\frac{11}{2}}$ *term of* $(\tfrac{1}{4}x^{-1}+x)^{\frac{1}{2}}$

809 Find numerical value correct to 5 significant figures.

1. $(1.01)^9$ 2. $(99)^4$ 3. $(103)^{\frac{1}{2}}$ 4. $(10)^{\frac{2}{3}}$ 5. $(1.03)^{-7}$

9 Exponential and Logarithmic Functions

Polynomials (Chapter 5) are a sum of one or more terms each of which is an *algebraic power function* x^b. The *variable x* is called the *base* and the *constant b* is called the *exponent*. For example:

(1) $x^5 - 4x^4 + x^3 - 13x^2 - 6x^1 + 6x^0 \rightarrow x^0 = 1$

Exponential function If, on the other hand, we have a function b^x in which the variable x appears as the exponent while positive real number constant b appears as the base, then we have the *exponential function* b^x.

(2) $y = b^x \rightarrow b > 1$ *and real*

Base b can be any number real or complex. However base b is real in this Chapter. Complex bases b is another subject.

The graph of the exponential function lies entirely above the x axis. Exponential function b^x is positive for every real value of x.

(3) *if* $x < 0 \rightarrow 0 < b^x < 1,$

    if $x = 0 \rightarrow \quad b^x = 1,$

    if $x > 0 \rightarrow \quad b^x > 1$

Corresponding to each real value if x there is one and only one value of y. The exponential function is a single valued function.

Examples Several examples of this new symbolism follow.

exponential form	logarithmic form	exponential form	logarithmic form
$64 = 4^3$	$\log_4 64 = 3$	$\frac{1}{32} = 8^{-\frac{5}{3}}$	$\log_8 \frac{1}{32} = -\frac{5}{3}$
$27 = 9^{\frac{3}{2}}$	$\log_9 27 = \frac{3}{2}$	$1 = 7^0$	$\log_7 1 = 0$

The Base e Exponential Function Leonhard Euler (1707-1783) defined the base e exponential function when he proved that sum of a power series solution to a first order differential equation is his number e raised to the $-\alpha x$ power ($e^{-\alpha x}$). He also proved the important property that the derivative of $e^{-\alpha x}$ is $-\alpha e^{-\alpha x}$ and calculated e=2.718281827....

Consider the equation of a parallel RC electric circuit.

$$0 = Gv(t) + C\frac{dv(t)}{dt} \qquad \Rightarrow \qquad -\frac{1}{RC}dt = \frac{dv}{v}$$

Integrate both sides using x as a dummy variable.

$$-\frac{1}{RC}\int_0^t dx = \int_0^t \frac{dv(x)}{v(x)} \rightarrow -\frac{1}{RC}(t-0) = \ln v(t) - \ln v(0)$$

$$-\frac{1}{RC}t = \ln\frac{v(t)}{v(0)} \qquad \Rightarrow \qquad v(t) = v(0)e^{-\frac{t}{RC}}$$

Plots of e^{-x} *and* $1-e^{-x}$ (the maximum value is 0.9 to clearly show the plot)

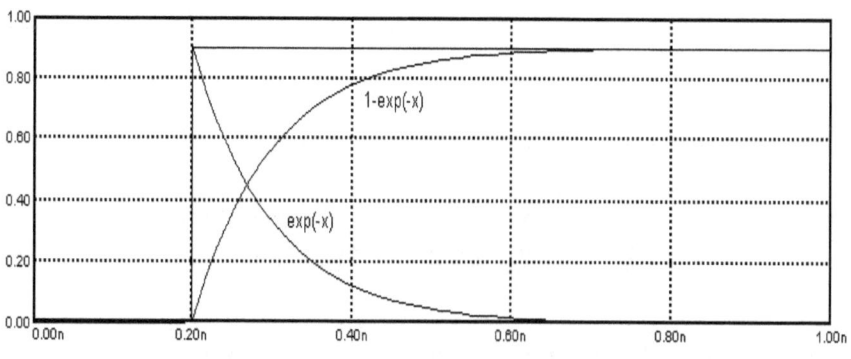

Problem 901 Let b=2. Plot $y=2^x$ when x = −3, −2, −1, 0, 1, 2, 3

Problem 902 Let b=10. Plot $y=10^x$ when x = −3, −2, −1, 0, 1, 2, 3

Problem 903 Let b= 1. Plot $y=1^x$ when x = −3, −2, −1, 0, 1, 2, 3.

Logarithmic function The logarithmic function is the inverse of the exponential function.

(4) $x = \log_b y \iff y = b^x \to b > 1$

Read this as "x is equal to the logarithm of y to the base b."

(5) *The exponent is* $x = \log_b y$ *and* $b^x = y$ *is the antilogarithm*

The base is b, the exponent/logarithm is x and the antilogarithm is y.

Bases most frequently used are 10 and e Although any positive number b greater than 1 may be taken as the base of a system of logarithms only two numbers are widely used in practice: 10 and Euler's e=2.71828...., which appears in the Calculus. The base must be a positive number greater than 1. Why? If base b=1, then $y=b^x=1^x=1$, which goes nowhere.

> The logarithm (log) *to the base 10* of a number is the exponent to which 10 must be raised to obtain the number.

Base 10 The log of 100 is 2, because $10^2=100$. The log of 0.001 is −3, because $10^{-3}=0.001$. The logarithms of most numbers are represented as decimals to so-many places. The log of 4.55 is 0.6580...., because $4.55=10^{0.6580}$.

> *Logarithms exist for all positive numbers, negative numbers and complex numbers. We only discuss the logarithm of positive numbers.*

A discussion about calculating with logarithms conveniently starts with logarithms of numbers from 1 to 10 whose values range from 0 to 1. The range's 0, 1 end values are the 0 and 1 exponents of 10 ($10^0=1$, $10^1=10$).

The whole number part of a logarithm is referred to as the *characteristic*, and the decimal part is referred to as the *mantissa*. For example, for log 4.55=0.6580 the 0 is the *characteristic* and .6580 is the *mantissa*.

Any number greater than 10 can be written as a number in the range 1 to 10 times a *positive* power of ten. We now show that the power of ten is the characteristic. For example when the power is 3.

(6) $4550 = 4.55 \times 10^3 = 10^{0.6580} \times 10^3 = 10^{3.6580}$ *and* $\log 4550 = 3.6580$

Algebra

And any number from 0 to 1 can be written as a number in the range 1 to 10 times a *negative* power of ten.

(7) $0.0455 = 4.55 \times 10^{-2} = 10^{0.6580} \times 10^{-2} = 10^{-1.3420}$ *and* $\log 0.0455 = -1.3420$

Gathering these results we can say
1 the logarithm of positive numbers greater than 1 is positive
2 the logarithm of 1 is zero
3 the logarithm of positive numbers less than 1 is negative.

Problem 904 Let b=2. Plot $\log_b 2^x$ when x = −3, −2, −1, 0, 1, 2, 3
Problem 905 Let b=10. Plot $\log_b 10^x$ when x = −3, −2, −1, 0, 1, 2, 3
Problem 906 Let b= 1. Plot $\log_b 1^x$ when x = −3, −2, −1, 0, 1, 2, 3.

9.1 Properties of Logarithms

Laws of exponents For example when x is an integer.

(8) $b^1 = b, \qquad b^2 = b \times b, \qquad b^3 = b \times b \times b$

Let b be any positive number, and x be any positive real number *not necessarily an integer*. Then the symbol b^x means the result of raising b to the power x. From this definition we have the following rules.

(9a) $b^x \cdot b^y = b^{x+y}$ *law for multiplication*

(9b) $\dfrac{b^x}{b^y} = b^{x-y}$ *law for division*

(9c) $(b^x)^y = b^{xy}$ *law for a power of a power*

The laws of exponents are valid for *all positive real numbers.*[1] (Not just integers.)

Negative Exponents

(10) *negative exponents* $y = b^{-x} = \dfrac{1}{b^x} = \dfrac{1}{m}$ *calculate* $m = b^x$, *then* $y = \dfrac{1}{m}$

For example

(11) $m = b^x = (0.7063)^{\frac{1}{6}} = 0.943695 \rightarrow y = \dfrac{1}{m} = 1.059665$

---

[1] The proof is beyond the scope of this text.

Properties of logarithms

1. The logarithm of a product is equal to the sum of the logarithms of its factors, all logarithms being taken to the same base.

let	$p = \log_b w, \quad q = \log_b z$
rewrite in exponential form	$w = b^p, \quad z = b^q$
multipy w by z	$wz = b^p b^q = b^{p+q}$
rewrite in logarithmic form	$\log_b wz = (p+q)\log_b b = p+q$
replace p and q by their values	$\log_b wz = \log_b w + \log_b z$

2. The logarithm of a quotient is equal to the logarithm of the dividend minus the logarithm of the divisor, all logarithms being taken to the same base.

let	$p = \log_b w, \quad q = \log_b z$
rewrite in exponential form	$w = b^p, \quad z = b^q$
divide w by z	$\dfrac{w}{z} = \dfrac{b^p}{b^q} = b^{p-q}$
rewrite in logarithmic form	$\log_b \dfrac{w}{z} = (p-q)\log_b b = p-q$
replace p and q by their logs	$\log_b \dfrac{w}{z} = \log_b w - \log_b z$

3. The logarithm of a power q of a number w is equal to the power q times the logarithm of w, all logarithms being taken to the same base.

let	$p = \log_b w$
rewrite in exponential form	$w = b^p$
raise to the qth power	$w^q = (b^p)^q = b^{pq}$
rewrite in logarithmic form	$\log_b w^q = qp\log_b b = qp$
replace p by its log	$\log_b w^q = q\log_b w$

Algebra

9.2 Solving Exponential and Logarithmic Equations

Exponential equations

(12) $\;2^{3x} = 4^{(x+1)}$

$3x\log 2 = (x+1)\log 4 = (x+1)\log 2^2 = 2(x+1)\log 2 \quad cancel \;\; \log 2$

$3x = 2(x+1) = 2x+2 \quad \Rightarrow \quad x = 2$

For some problems you need a calculator (or a table of logs).

(13) $\;n = 4.92^{5.368}$

$\log n = 5.368 \log 4.92 = 5.368 \times 0.691965103 = 3.714468672$

$\log n = 3 + 0.714468672 = \log 10^3 + \log 5.181657126 \;\;\rightarrow\;\; n = 5.181657126 \times 10^3$

(14) $\;23.45 = b^{\frac{1}{7}}$

$\log 23.45 = \dfrac{1}{7}\log b \;\;\Rightarrow\;\; \log b = 7\log 23.45 = 7 \times 1.370143 = 9.59099993$

$\log b = 9 + 0.59099993 = \log 10^9 + \log 3.899424446 \;\;\rightarrow\;\; b = 3.899419229 \times 10^9$

(15) $\;801.2 = 14.56^d$

$\log 801.2 = d\log 14.56 \;\;\Rightarrow\;\; d = \dfrac{\log 801.2}{\log 14.56} = \dfrac{2.903740941}{1.163161375} = 2.496421394$

$d = 2.496421394$

Logarithmic equations Sometimes the numbers fall out.

(16) $\;$ *solve for n* $\;\log_5 n = -2 \;\;\rightarrow\;\; n = 5^{-2} = \dfrac{1}{25}$

(17) $\;$ *solve for b* $\;\log_b 16 = \dfrac{2}{3} \;\;\rightarrow\;\; 16 = b^{\frac{2}{3}} \;\;\Rightarrow\;\; b = 16^{\frac{3}{2}} = \left(16^{\frac{1}{2}}\right)^3 = 4^3 = 64$

(18) $\;2\log x - \log(30-2x) = 1 \;\;\rightarrow\;\; \log x^2 - \log(30-2x) = 1$

$$\log \frac{x^2}{30-2x} = 1$$

$$\frac{x^2}{30-2x} = 10^1 = 10$$

$$x^2 + 20x - 300 = 0 \quad \Rightarrow \quad x = -30, \;\; x = 10$$

9.3 Logarithms to bases to other than 10

To find the relation between ln y (base e) and log y (base 10) start with

(20a) $y = \ln x$ if $e^y = x$ then $e^{\ln x} = x$

(20b) $y = \log x$ if $10^y = x$ then $10^{\log x} = x$

(21a) *from 20a we have* $p = e^{\ln p}$

(21b) $\log p = \log(e^{\ln p})$ $\Rightarrow$ $\log p = \ln p \times \log(e)$

(21c) $\log e = \log 2.71828 = 0.43429$

(21d) $\log p = 0.43429 \ln p$

In general the relation between logarithms of the same number y to different bases b is given by the following theorem.

(22)   *Theorem* $\log_a y = \dfrac{\log_b y}{\log_b a}$

Proof   let $x = \log_a y$ $\rightarrow$ $y = a^x$

$\rightarrow$ *take log of both sides* $\log_b y = x\log_b a$ $\rightarrow$ $x = \log_a y = \dfrac{\log_b y}{\log_b a}$

Useful Relations

(23)   *Let* $y = b$ $\log_a b = \dfrac{\log_b b}{\log_b a} = \dfrac{1}{\log_b a} = \dfrac{1}{\log_b a}$

Important specific cases

(24)   *Bases b = 10 and a = e* $\log_e y = \dfrac{\log_{10} y}{\log_{10} e}$ $\Rightarrow$ $\ln y = \dfrac{\log y}{\log e}$

(25)   *Bases a = 10 and b = e* $\log_{10} y = \dfrac{\log_e y}{\log_e 10}$ $\Rightarrow$ $\log y = \dfrac{\ln y}{\ln 10}$

$\log e = 0.4343$ *and* $\ln 10 = 2.3026$ *and* $0.4343 = \dfrac{1}{2.3026}$

Algebra

Problems 9

Convert exponential form y = b^x to logarithmic form x = log$_b$ y.

1. $3^4 = 81$ 2. $8^{\frac{1}{3}} = 2$ 3. $2^3 = 8$ 4. $10^4 = 10000$ 5. $10^{-2} = \frac{1}{100}$

6. $\left(\dfrac{1}{4}\right)^3 = \dfrac{1}{64}$ 7. $7^x = y$ 8. $b^3 = y$ 9. $10^x = y$ 10. $b^x = 27$

Convert logarithmic form x = log$_b$ y to exponential form y = b^x.

11. $5 = \log_2 32$ 12. $\dfrac{1}{4} = \log_{16} 2$ 13. $4 = \log_2 16$

14. $6 = \log_{10} 1000000$ 15. $-3 = \log_2 \dfrac{1}{8}$

16. $-3 = -\log_2 8$ 17. $z = \log_{10} y$ 18. $5 = \log_3 243$

19. $-3 = \log_{10} 0.001$ 20. $x = \log_3 81$

Find log$_{10}$ of the following numbers.

21. 100   22. 0.01   23. 1000   24. 1   25. 0.001

26. 100000   27. 0.00001   28. 10   29. 0.1   30. 0.001

Solve for x.

31. $\log_{10} x = 5$ 32. $\log_x 16 = 4$ 33. $\log_2 x = 5$ 34. $\log_4 64 = x$

35. $\log_{16} x = \frac{3}{2}$ 36. $\log_x 27 = \frac{3}{4}$ 37. $\log_{25} 625 = x$ 38. $\log_4 x = \frac{5}{2}$

Find characteristic of log$_{10}$ y for following y.

39. 7.234   40. 72.34   41. 0.7234   42. 72340   43. 7234×10^4

44. 0.007234   45. 72.34×10^{-6} 46. 723400   47. 0.72340×10^{-4}

If log$_{10}$ y = 0.69897, then y = 5. Use laws for log xy and log x/y to find log of the following numbers.

48. $\log 10y$ 49. $\log \dfrac{y}{10}$ 50. $\log 100y$ 51. $\log 1000y$ 52. $\log \dfrac{y}{1000}$

53. $\log 2 (hint\ 2 = \frac{10}{5})$ 54. $\log \frac{2}{10}$ 55. $\log 200$ 56. $\log \frac{1}{2}$ 57. $\log \frac{100}{2}$

Log 10 = 1, Log 5 = 0.69897, 1−0.69897 = 0.30103. Find the antilog of the following numbers.

58. 1−0.69897   59. 2.69897   60. −1+0.30103   61. 2+3.69897

62. −1+0.69897   63. 4.30103   64. 2.30103−1.69897   65. 2.69897−1.30103

Find the value of these expressions.

66. $\log 10^3$ 67. $\log(0.01)^4$ 68. $\log(0.001)^3$ 69. $\log 5^3$ 70. $\log 2^4$

Expand as algebraic sum of terms.

71. $\log 3^2 7^3 5^7$ 72. $\log 9^{-1} 7^2$ 73. $\log 4^{\frac{1}{2}} 8^{\frac{1}{3}}$ 74. $\log 5^2 4^3$ 75. $3\log 5^2 7$

10 Partial Fractions

In many problems a *rational function*, the ratio of two polynomials $N(p)/D(p)$, is decomposed into a sum of fractions with denominators of lower degree. Each fraction in the sum is referred to as a *partial fraction*. This process is the inverse of a process that adds fractions.

Given a ratio of two polynomials in the variable p. We can always divide the denominator into the numerator so that the remainder $F(p)$ is a proper rational fraction, which means the degree of $N(p)$ is less than the degree of $D(p)$. For example:

$$(1) \quad \text{if } G(p) = \frac{p^4 + 5p^3 + 8p^2 + 3p + 6}{p^3 + 4p^2 + 3p} = p + 1 + \frac{p^2 + 6}{p^3 + 4p^2 + 3p}$$

$$(2) \quad \text{let } F(p) = \frac{p^2 + 6}{p^3 + 4p^2 + 3p} = \frac{N(p)}{D(p)} \quad (a \text{ proper fraction})$$

There are two problems to solve:
1. Find the roots of $D(p)$ (we assume you know how to do this, see Chapter 5).

2. Expand $F(p)$ into a sum of partial fraction terms.

We will show by examples the processes for partial fraction expansions.

The basis for decomposing a proper fraction into a sum of partial fractions is the *Theorem* in the side bar, the proof[1] of which is beyond the scope of this text.

The methods decomposing a ratio of two polynomials $N(p)/D(p)$ into a sum of fractions are based on the fact coefficients of similar terms of two equal polynomials are equal.

---

[1] W. L. Ferrar *Higher Algebra* Oxford Press ISBN 0198325061

Theorem   *Any proper fraction N(p)/D(p) whose numerator and denominator are polynomials in p may be decomposed into an algebraic sum of partial fractions of the types listed here.*

1. If any linear factor, such as (ap+b), occurs *once* as a factor of the denominator D(p) of the given function there will correspond to that factor a partial fraction with the form

$$\frac{A}{ap+b}$$

2. If any linear factor, such as (ap+b), occurs k times as a factor of the denominator of D(p) of the given function there will correspond to that factor the sum of k partial fractions with the form

$$\frac{A_1}{ap+b} + \frac{A_2}{(ap+b)^2} + \cdots + \frac{A_k}{(ap+b)^k} \quad \text{where the A's are constants and } A_k \neq 0$$

3. If any quadratic factor, such as (ap²+bp+c), occurs *once* as a factor of the denominator of D(p) of the given function there will correspond to that factor a partial fraction with the form

$$\frac{Ap+B}{ap^2+bp+c}$$

4. If any quadratic factor, such as (ap²+bp+c), occurs k times as a factor of the denominator of D(p) of the given function there will correspond to that factor the sum of k partial fractions with the form

$$\frac{A_1 p + B_1}{ap^2+bp+c} + \frac{A_2 p + B_2}{(ap^2+bp+c)^2} + \cdots + \frac{A_k p + B_k}{(ap^2+bp+c)^k}$$

Note: In every case the number of constants in the numerator of a partial fraction equals the degree of the denominator n of the fraction. The numerator is then a polynomial of degree n−1.

10.1 Sums of Fractions

We know from arithmetic that it is always possible to express the sum or difference of a number of fractions as one fraction. The same can be said for fractions with polynomials. Consider these examples of proper fraction sums implemented by using the trick a/a=1.

create common denominator by multiplying by a/a and b/b

(3) $f(p) = \dfrac{1}{p+1} - \dfrac{1}{p+2} = \dfrac{1}{p+1}\dfrac{a}{a} - \dfrac{1}{p+2}\dfrac{b}{b} \quad \rightarrow \quad a = p+2 \text{ and } b = p+1$

$\quad = \dfrac{1}{p+1}\cdot\dfrac{p+2}{p+2} - \dfrac{1}{p+2}\cdot\dfrac{p+1}{p+1}$

fraction manipulations :

$\dfrac{n_1}{d} - \dfrac{n_2}{d} = \dfrac{1}{d}\dfrac{n_1}{1} - \dfrac{1}{d}\dfrac{n_2}{1} = \dfrac{1}{d}\left(\dfrac{n_1}{1} - \dfrac{n_2}{1}\right) = \dfrac{1}{d}\left(n_1 - n_2\right) = \dfrac{1}{d}\left(\dfrac{n_1-n_2}{1}\right) = \dfrac{n_1-n_2}{d}$

$f(p) = \dfrac{(p+2)-(p+1)}{(p+1)(p+2)} = \dfrac{p+2-p-1}{(p+1)(p+2)} = \dfrac{1}{(p+1)(p+2)}$

(4) $f(p) = \dfrac{1}{p+1} - \dfrac{2}{p+2} = \dfrac{(p+2)-2(p+1)}{(p+1)(p+2)} = -\dfrac{p}{(p+1)(p+2)}$

(5) $f(p) = \dfrac{p+1}{p^2+1} + \dfrac{2p}{p^2+2} = \dfrac{(p+1)(p^2+2)+2p(p^2+1)}{(p^2+1)(p^2+2)}$

$\quad = \dfrac{p(p^2+2)+1(p^2+2)+2p(p^2+1)}{(p^2+1)(p^2+2)}$

$\quad = \dfrac{p^3+2p+p^2+2+2p^3+2p}{(p^2+1)(p^2+2)} = \dfrac{3p^3+p^2+4p+2}{(p^2+1)(p^2+2)}$

Emphasis: Observe that the sums are also proper fractions. The degree of the numerator is less than the degree of the denominator.

In many branches of mathematics the reverse operation needs to be executed. Given a fraction whose denominator is factored, express the fraction as an algebraic sum of partial fractions, where each partial fraction is a proper fraction.

10.2 Linear Factors of order 1

When the denominator of a proper fraction F(p) can be resolved into real linear factors, all of which are distinct.

Consequently each partial fraction has to be a proper fraction, where A_1, A_2, and A_3 are real constants so that the degree of the numerators less than the degree of the denominators. Reference Theorem part 1.

(6) $\quad F(p) = \dfrac{p^2 + 6}{p^3 + 4p^2 + 3p}$

(7) $\quad \dfrac{p^2 + 6}{p(p+1)(p+3)} = \dfrac{A_1}{p} + \dfrac{A_2}{p+1} + \dfrac{A_3}{p+3}$

cross multiply :

(8) $\quad p^2 + 6 = A_1(p+1)(p+3) + A_2 p(p+3) + A_3 p(p+1)$

Do not simplify by multiplying out. It is easier to substitute selected values of p.

(9) *if* $p = 0$ *then*

$\quad 0 + 6 = A_1(0+1)(0+3) + A_2 0(0+3) + A_3 0(0+1)$

$\quad 6 = 3A_1 \quad \Rightarrow \quad A_1 = 2$

(10) *if* $p = -1$ *then*

$\quad 1 + 6 = A_1(-1+1)(-1+3) + A_2(-1)(-1+3) + A_3(-1)(-1+1)$

$\quad 7 = -2A_2 \quad \Rightarrow \quad A_2 = -\dfrac{7}{2}$

(11) *if* $p = -3$ *then*

$\quad 9 + 6 = A_1(-3+1)(-3+3) + A_2(-3)(-3+3) + A_3(-3)(-3+1)$

$\quad 15 = 6A_3 \quad \Rightarrow \quad A_3 = \dfrac{5}{2}$

Check

(12) $\quad F(p) = \dfrac{A_1}{p} + \dfrac{A_2}{p+1} + \dfrac{A_3}{p+3} = \dfrac{2}{p} + \dfrac{-7/2}{p+1} + \dfrac{5/2}{p+3}$

$\quad 2F(p) = \dfrac{4}{p} - \dfrac{7}{p+1} + \dfrac{5}{p+3} = \dfrac{4(p+1)(p+3) - 7p(p+3) + 5p(p+1)}{p(p+1)(p+3)}$

$\quad F(p) = \dfrac{1}{2} \dfrac{(4-7+5)p^2 + (16-21+5)p + 12}{p(p+1)(p+3)} = \dfrac{p^2 + 6}{p(p+1)(p+3)} \quad qed$

10.3 Linear Factors of order k

When the denominator of a proper fraction F(p) can be resolved into real linear factors, some of which are repeated.

As before each partial fraction has to be a proper fraction.

Each higher order factor of order n requires a sum of terms. There is one term for each power of p, or p+1, from 1 to k. Reference Theorem part 2.

(13) $F(p) = \dfrac{5p^3 - 6p - 3}{p^3(p+1)^2} = \dfrac{A_1}{p} + \dfrac{A_2}{p^2} + \dfrac{A_3}{p^3} + \dfrac{A_4}{p+1} + \dfrac{A_5}{(p+1)^2}$

cross multiply

(14a) $5p^3 - 6p - 3 = A_1 p^2 (p+1)^2 + A_2 p(p+1)^2$

$$+ A_3(p+1)^2 + A_4 p^3(p+1) + A_5 p^3$$

(14b) $5p^3 - 6p - 3 = A_1(p^4 + 2p^3 + p^2) + A_2(p^3 + 2p^2 + p)$

$$+ A_3(p^2 + 2p + 1) + A_4(p^4 + p^3) + A_5 p^3$$

(15) $5p^3 - 6p - 3 = p^4(A_1 + A_4) + p^3(2A_1 + A_2 + A_4 + A_5) + p^2(A_1 + 2A_2 + A_3) +$

$$+ p(A_2 + 2A_3) + 1(A_3)$$

From 14a

Let $p = 0$ $-3 = 0 + 0 + 0 + A_3 + 0 + 0$ $\Rightarrow$ $A_3 = -3$

Let $p = -1$ $-5 + 6 - 3 = 0 + 0 + 0 + 0 + 0 - A_5$ $\Rightarrow$ $A_5 = 2$

From 15

equate coefficients of terms

$p^4 : 0 = A_1 + A_4$ $p^3 : 5 = 2A_1 + A_2 + A_4 + A_5$ $p^2 : 0 = A_1 + 2A_2 + A_3$

$p : -6 = A_2 + 2A_3$ $1 : -3 = A_3$

(16) $A_3 = -3$, $A_2 = 0$, $A_1 = 3$, $A_4 = -3$, $A_5 = 2$

(17) $F(p) = \dfrac{5p^3 - 6p - 3}{p^3(p+1)^2} = \dfrac{3}{p} + \dfrac{0}{p^2} - \dfrac{3}{p^3} - \dfrac{3}{p+1} + \dfrac{2}{(p+1)^2}$

Algebra

Check

(18) $F(p) = \dfrac{A_1}{p} + \dfrac{A_2}{p^2} + \dfrac{A_3}{p^3} + \dfrac{A_4}{p+1} + \dfrac{A_5}{(p+1)^2} = \dfrac{3}{p} + \dfrac{0}{p^2} - \dfrac{3}{p^3} - \dfrac{3}{p+1} + \dfrac{2}{(p+1)^2}$

$= \dfrac{3p^2(p+1)^2 - 3(p+1)^2 - 3p^3(p+1) + 2p^3}{p^3(p+1)^2}$

$= \dfrac{3(p+1)[p^2(p+1)-(p+1)-p^3]+2p^3}{p^3(p+1)^2} = \dfrac{3(p+1)[p^3+p^2-p-1-p^3]+2p^3}{p^3(p+1)^2}$

$= \dfrac{(3p+3)[p^2-p-1]+2p^3}{p^3(p+1)^2} = \dfrac{[3p^3+3p^2-3p^2-3p-3p-3]+2p^3}{p^3(p+1)^2}$

$= \dfrac{5p^3-6p-3}{p^3(p+1)^2}$ *qed*

10.4 Quadratic Factors of order 1

When the denominator of a proper fraction F(p) contains one quadratic factor, but no repeated quadratic factor. Reference Theorem part 3.

(19) $F(p) = \dfrac{16}{p(p^2+5p+2)}$

(20) $\dfrac{16}{p(p^2+5p+2)} = \dfrac{A_1}{p} + \dfrac{A_2 p + B_2}{p^2+5p+2}$ *now cross multiply*

$16 = A_1(p^2+5p+2) + (A_2 p + B_2)p$

(21) $16 = A_1 p^2 + 5A_1 p + 2A_1 + A_2 p^2 + B_2 p$

equate coefficients of terms

$p^2 : 0 = A_1 + A_2 \qquad p : 0 = 5A_1 + B_2 \qquad 1 : 16 = 2A_1$

(22) $A_1 = 8, \; A_2 = -8, \; B_2 = -40$

(23) $F(p) = \dfrac{16}{p(p^2+5p+2)} = \dfrac{8}{p} - \dfrac{8p+40}{p^2+5p+2}$

Check

(24) $F(p) = \dfrac{A_1}{p} + \dfrac{A_2 p + B_2}{p^2+5p+2} = \dfrac{8}{p} - \dfrac{8p+40}{p^2+5p+2} = \dfrac{8(p^2+5p+2)-p(8p+40)}{p(p^2+5p+2)}$

$= \dfrac{8p^2+40p+16-8p^2-40p}{p(p^2+5p+2)} = \dfrac{16}{p(p^2+5p+2)}$ *qed*

10.5 Quadratic Factors of order k

When the denominator of a proper fraction F(p) contains one quadratic repeated factor. Reference Theorem part 4.

(25) $F(p) = \dfrac{16}{p(p^2 + 5p + 2)^2}$

(26) $\dfrac{16}{p(p^2 + 5p + 2)^2} = \dfrac{A_1}{p} + \dfrac{A_2 p + B_2}{p^2 + 5p + 2} + \dfrac{A_3 p + B_3}{(p^2 + 5p + 2)^2}$

Now cross multiply.

$$16 = A_1(p^2 + 5p + 2)^2 + (A_2 p + B_2)p(p^2 + 5p + 2) + (A_3 p + B_3)p$$

(27) $16 = A_1(p^4 + 10p^3 + 29p^2 + 20p + 4)$

$$+ A_2(p^4 + 5p^3 + 2p^2) + B_2(p^3 + 5p^2 + 2p) + A_3 p^2 + B_3 p$$

$$= p^4(A_1 + A_2) + p^3(10A_1 + 5A_2 + B_2)$$

$$+ p^2(29A_1 + 2A_2 + 5B_2 + A_3) + p(20A_1 + 2B_2 + B_3) + 1(4A_1)$$

equate coefficients of terms

$p^4 : 0 = A_1 + A_2$ $\qquad\qquad$ $p^3 : 0 = 10A_1 + 5A_2 + B_2$

$p^2 : 0 = 29A_1 + 2A_2 + 5B_2 + A_3$ $\qquad$ $p: 0 = 20A_1 + 2B_2 + B_3$ $\quad$ $1: 16 = 4A_1$

(28) $A_1 = 4,\ A_2 = -4,\ B_2 = -20,\ B_3 = -40,\ A_3 = -8$

Check

(29) $F(p) = \dfrac{A_1}{p} + \dfrac{A_2 p + B_2}{p^2 + 5p + 2} + \dfrac{A_3 p + B_3}{(p^2 + 5p + 2)^2}$

$$= \dfrac{4}{p} - \dfrac{4p + 20}{p^2 + 5p + 2} - \dfrac{8p + 40}{(p^2 + 5p + 2)^2}$$

$$= \dfrac{4(p^2 + 5p + 2)^2 - p(4p + 20)(p^2 + 5p + 2) - p(8p - 40)}{p(p^2 + 5p + 2)^2}$$

$$= \dfrac{(p^2 + 5p + 2)(4p^2 + 20p + 8 - 4p^2 - 20p) - 8p^2 - 40p}{p(p^2 + 5p + 2)^2}$$

$$= \dfrac{(p^2 + 5p + 2)(8) - 8p^2 - 40p}{p(p^2 + 5p + 2)^2} = \dfrac{16}{p(p^2 + 5p + 2)^2} \quad qed$$

Problems 10

Convert into a sum of partial fractions as shown. Show all steps

1 $\dfrac{p-13}{(p-3)(p+2)} = \dfrac{3}{p+2} - \dfrac{2}{p-3}$

2 $\dfrac{x+2}{2x^2 - x} = \dfrac{5}{2x-1} - \dfrac{2}{x}$

3 $\dfrac{3y^2 + 4y - 15}{(y-3)(y-1)(y+1)} = \dfrac{3}{y-3} + \dfrac{2}{y-1} - \dfrac{2}{y+1}$

4 $\dfrac{z^2 - 2z + 16}{z^3 - 4z} = -\dfrac{4}{z} + \dfrac{2}{z-2} + \dfrac{3}{z+2}$

5 $\dfrac{x^2 - 11x - 6}{2x^3 - x^2 - 8x + 4} = \dfrac{3}{2x-1} - \dfrac{2}{x-2} + \dfrac{1}{x+2}$

6 $\dfrac{x^2}{(x+1)^2(x-1)} = \dfrac{1}{4(x-1)} + \dfrac{3}{4(x+1)} - \dfrac{1}{2(x+1)^2}$

7 $\dfrac{15 - 12x}{(x-2)^2(2x-1)^2} = \dfrac{4}{(2x-1)^2} - \dfrac{1}{(x-2)^2}$

8 $\dfrac{x^2 - x + 13}{(x+1)(x^2 + 4)} = \dfrac{3}{x+1} + \dfrac{1 - 2x}{x^2 + 4}$

9 $\dfrac{x^4 - 2x^3 + 2x^2 - 2x + 2}{x^3 - x^2 + x - 1} = x - 1 + \dfrac{1}{2}\left(\dfrac{1}{(x-1)} - \dfrac{x+1}{(x^2 + 1)} \right)$

11 Matrix Algebra

A matrix is an array of r×c numbers, real or complex, arranged in r rows and c columns. Matrices allow one to write and process equations efficiently. Furthermore, in many problems, the matrix format makes the next step easier to perceive. This will become clear as we proceed. The world says a matrix has rows and columns. This matrix has 2 rows and 3 columns. D is a 2×3 matrix.

(1) $D = D_{row \times column} = D_{2 \times 3} = \begin{bmatrix} 3 & 7 & 9 \\ 4 & 5 & -1 \end{bmatrix}$

(2a) *rows*   (3 7 9)   (4 5 −1)

(2b) *columns* $\begin{pmatrix} 3 \\ 4 \end{pmatrix} \begin{pmatrix} 7 \\ 5 \end{pmatrix} \begin{pmatrix} 9 \\ -1 \end{pmatrix}$

Matrix equation AX=B represents two equations.
(3a) $2x + 9y = -7$ (2b) $5x + 3y = 1$

(3b) $\begin{bmatrix} 2 & 9 \\ 5 & 3 \end{bmatrix} \times \begin{bmatrix} x \\ y \end{bmatrix} = \begin{bmatrix} -7 \\ 1 \end{bmatrix}$

(3c) $AX = B$

The zero, null, 2×2 matrix is (4) $\begin{bmatrix} 0 & 0 \\ 0 & 0 \end{bmatrix}$

11.1 Matrix Addition and Subtraction

Add matrices by adding corresponding elements. Subtract by replacing + by −. Observe that addition and subtraction requires *conformable* matrices. For example consider 2×3 and 3×2 matrices.

(5a) $A + B = \begin{bmatrix} a_{11} & a_{12} & a_{13} \\ a_{21} & a_{22} & a_{23} \end{bmatrix} + \begin{bmatrix} b_{11} & b_{12} & b_{13} \\ b_{21} & b_{22} & b_{23} \end{bmatrix} = \begin{bmatrix} a_{11}+b_{11} & a_{12}+b_{12} & a_{13}+b_{13} \\ a_{21}+b_{21} & a_{22}+b_{22} & a_{23}+b_{23} \end{bmatrix}$

(5b) $C + D = \begin{bmatrix} a_{11} & a_{12} \\ a_{21} & a_{22} \\ a_{31} & a_{32} \end{bmatrix} + \begin{bmatrix} b_{11} & b_{12} \\ b_{21} & b_{22} \\ b_{31} & b_{32} \end{bmatrix} = \begin{bmatrix} a_{11}+b_{11} & a_{12}+b_{12} \\ a_{21}+b_{21} & a_{22}+b_{22} \\ a_{31}+b_{31} & a_{32}+b_{32} \end{bmatrix}$

Two matrices are conformable for addition when each has the same number of rows, and each has the same number of columns.

Algebra

11.2 Matrix Multiplication

Any number q times matrix A multiplies each element of A q times. Let q=3.

$$(6a) \quad 3A = A + A + A = \begin{bmatrix} a_{11} & a_{12} \\ a_{21} & a_{22} \end{bmatrix} + \begin{bmatrix} a_{11} & a_{12} \\ a_{21} & a_{22} \end{bmatrix} + \begin{bmatrix} a_{11} & a_{12} \\ a_{21} & a_{22} \end{bmatrix} = \begin{bmatrix} 3a_{11} & 3a_{12} \\ 3a_{21} & 3a_{22} \end{bmatrix}$$

$$(6b) \quad qA = \begin{bmatrix} qa_{11} & qa_{12} \\ qa_{21} & qa_{22} \end{bmatrix}$$

The matrix $-B$ is a matrix whose elements are those of B multiplied by -1. I.e. $q = -1$. We can demonstrate this by subtraction $0 - B = -B$.

$$(7a) \quad 0 - B = \begin{bmatrix} 0 & 0 \\ 0 & 0 \end{bmatrix} - \begin{bmatrix} b_{11} & b_{12} \\ b_{21} & b_{22} \end{bmatrix} = \begin{bmatrix} 0 - b_{11} & 0 - b_{12} \\ 0 - b_{21} & 0 - b_{22} \end{bmatrix} = \begin{bmatrix} -b_{11} & -b_{12} \\ -b_{21} & -b_{22} \end{bmatrix} = -B$$

$$(7b) \quad \text{If } B = \begin{bmatrix} b_{11} & b_{12} \\ b_{21} & b_{22} \end{bmatrix} \text{ then } -B = -1 \times B = \begin{bmatrix} -b_{11} & -b_{12} \\ -b_{21} & -b_{22} \end{bmatrix}$$

The dot, scalar, or inner product of two numbers is a guide to matrix multiplication.

(8a) x has components $x_1 \ x_2 \ x_3 \ .... x_n$

(8b) y has components $y_1 \ y_2 \ y_3 \ .... y_n$

(8c) then the dot product $x \cdot y = x_1 y_1 + x_2 y_2 + x_3 y_3 + ... + x_n y_n$

Must be conformable for multiplication One way to understand why matrices must be conformable is to try and multiply non-conformable matrices such as A and B.

$$(9) \quad \text{if } A = \begin{bmatrix} a_{11} \\ a_{21} \end{bmatrix} \text{ and } B = \begin{bmatrix} b_{11} & b_{12} \\ b_{21} & b_{22} \end{bmatrix} \text{ then } AB \text{ cannot be defined}$$

However the product BA is conformable for multiplication. Observe how B row 1 and A column 1 form a dot product, and how B row 2 and A column 1 form a dot product

$$(10) \quad BA = \begin{bmatrix} b_{11} & b_{12} \\ b_{21} & b_{22} \end{bmatrix} \times \begin{bmatrix} a_{11} \\ a_{21} \end{bmatrix} = \begin{bmatrix} b_{11}a_{11} + b_{12}a_{21} \\ b_{21}a_{11} + b_{22}a_{21} \end{bmatrix}$$

An elementary view of matrix multiplication

$$(11a) \quad \begin{bmatrix} 1 & 2 \\ - & - \end{bmatrix} \times \begin{bmatrix} 4 & - \\ 6 & - \end{bmatrix} = \begin{bmatrix} 1 \times 4 + 2 \times 6 & - \\ - & - \end{bmatrix} = \begin{bmatrix} 16 & - \\ - & - \end{bmatrix}$$

$$(11b) \quad \begin{bmatrix} - & - \\ 3 & 4 \end{bmatrix} \times \begin{bmatrix} 4 & - \\ 6 & - \end{bmatrix} = \begin{bmatrix} - & - \\ 3 \times 4 + 4 \times 6 & - \end{bmatrix} = \begin{bmatrix} - & - \\ 36 & - \end{bmatrix}$$

$$(11c) \quad \begin{bmatrix} 1 & 2 \\ - & - \end{bmatrix} \times \begin{bmatrix} - & 5 \\ - & 7 \end{bmatrix} = \begin{bmatrix} - & 1 \times 5 + 2 \times 7 \\ - & - \end{bmatrix} = \begin{bmatrix} - & 19 \\ - & - \end{bmatrix}$$

$$(11d) \quad \begin{bmatrix} - & - \\ 3 & 4 \end{bmatrix} \times \begin{bmatrix} - & 5 \\ - & 7 \end{bmatrix} = \begin{bmatrix} - & - \\ - & 3 \times 5 + 4 \times 7 \end{bmatrix} = \begin{bmatrix} - & - \\ - & 43 \end{bmatrix}$$

Linear substitution is also a guide to matrix multiplication. Consider the equations where a's and b's are constants.

$(12a) \quad x_1 = a_{11}y_1 + a_{12}y_2$ $\qquad$ $(13a) \quad y_1 = b_{11}z_1 + b_{12}z_2$

$(12b) \quad x_2 = a_{21}y_1 + a_{22}y_2$ $\qquad$ $(13b) \quad y_2 = b_{21}z_1 + b_{22}z_2$

$(12c) \quad X = AY$ $\qquad\qquad\qquad$ $(13c) \quad Y = BZ$

$(14a) \quad x_1 = a_{11}(b_{11}z_1 + b_{12}z_2) + a_{12}(b_{21}z_1 + b_{22}z_2)$

$(14b) \quad x_2 = a_{21}(b_{11}z_1 + b_{12}z_2) + a_{22}(b_{21}z_1 + b_{22}z_2)$

$(15a) \quad x_1 = (a_{11}b_{11} + a_{12}b_{21})z_1 + (a_{11}b_{12} + a_{12}b_{22})z_2$

$(15b) \quad x_2 = (a_{21}b_{11} + a_{22}b_{21})z_1 + (a_{21}b_{12} + a_{22}b_{22})z_2$

$(15c) \quad X = ABZ$

> *Two matrices A and B are conformable for multiplication as AB when number of columns in A equals the number of rows in B. In turn AB and Z are conformable for multiplication.*

The subscripts are row and column numbers. The first number is a row number. The second number is a column number.

Two matrices A, B are equal, and we write A=B, when the matrices are conformable and when each element of A equals the corresponding element of B.

Algebra

Matrices may or may not commute Matrix AB may not equal matrix BA. For example.

(16a) $\quad AB = \begin{bmatrix} 1 & 2 \\ 1 & 2 \end{bmatrix} \times \begin{bmatrix} 2 & 1 \\ 2 & 1 \end{bmatrix} = \begin{bmatrix} 6 & 3 \\ 6 & 3 \end{bmatrix}$ $\quad$ (16b) $\quad BA = \begin{bmatrix} 2 & 1 \\ 2 & 1 \end{bmatrix} \times \begin{bmatrix} 1 & 2 \\ 1 & 2 \end{bmatrix} = \begin{bmatrix} 3 & 6 \\ 3 & 6 \end{bmatrix}$

The unit matrix commutes The square matrix of order n that has ones in its leading diagonal and zeros elsewhere is referred to as the *unit matrix* of order n. The unit matrix symbol is I.

(17) $\quad I = I_{5\times 5} = \begin{bmatrix} 1 & 0 & 0 & 0 & 0 \\ 0 & 1 & 0 & 0 & 0 \\ 0 & 0 & 1 & 0 & 0 \\ 0 & 0 & 0 & 1 & 0 \\ 0 & 0 & 0 & 0 & 1 \end{bmatrix}$

(18a) $\quad IA = AI = A$ $\qquad$ (18b) $\quad I = I^2 = I^3 = ...$

Distributive and associative laws for multiplication apply to matrices

(19a) $\quad (A+B)C = AC + BC$ $\qquad$ (19b) $\quad (AB)C = A(BC)$

The Division Law The division law in ordinary algebra states that when the product xy=0, either x or y is zero, or both must be zero. This law *does not apply* to matrix products. The product AB may equal zero, however this does not imply A or B are the null matrix (Problem 1102).

Cancellation If ab=ac, a≠0, in the algebra of numbers, then we may cancel a on both sides of = so that b=c. However this may not be possible in a matrix equation such as AB=AC where B≠C (Problem 1103)

Problem 1101 Let A be a 2×2 matrix. Show that 0×A=A×0=0.

Problem 1102 $A = \begin{bmatrix} a & b \\ 0 & 0 \end{bmatrix}$ $\quad B = \begin{bmatrix} b & 2b \\ -a & -2a \end{bmatrix}$ Show that AB=0, BA≠0.

Problem 1103 $A = \begin{bmatrix} 0 & 0 \\ 0 & 1 \end{bmatrix}$ $\quad B = \begin{bmatrix} 1 & 1 \\ 1 & 1 \end{bmatrix}$ $\quad C = \begin{bmatrix} 0 & 0 \\ 1 & 1 \end{bmatrix} = D$

Show that AB=AC=D yet B≠C.

11.3 Related Matrices

Transpose of a matrix A^T Rows become columns or vice versa.

(20) $A = \begin{bmatrix} a_{11} & a_{12} & a_{13} \\ a_{21} & a_{22} & a_{23} \end{bmatrix}$ $\qquad A^T = \begin{bmatrix} a_{11} & a_{21} \\ a_{12} & a_{22} \\ a_{13} & a_{23} \end{bmatrix}$

(21a) $(AB)^T = B^T A^T$ $\qquad$ (21b) $(ABC)^T = C^T B^T A^T$

Reciprocal or Inverse of a matrix A^{-1} How to calculate the inverse is demonstrated in 11.5.

(22a) $AA^{-1} = I$ $\qquad$ (22b) $A^{-1}A = I$

(23a) $(A^{-1})^{-1} = A$ $\quad$ (23b) $(A^{-1}B^{-1})^{-1} = BA$

(24a) $(AB)^{-1} = B^{-1}A^{-1}$ $\qquad$ (24b) $(ABC)^{-1} = C^{-1}B^{-1}A^{-1}$

Reciprocating and Transposing commute

(25) $(A^T)^{-1} = (A^{-1})^T$

Positive and Negative Integer Matrix Exponents A^r

(26) $A^2 = AA$ $\quad A^3 = AAA$ $\quad A^{-2} = A^{-1}A^{-1}$ $\quad$ *and so forth*

(27) $A^r A^s = A^{r+s}$ $\quad (A^{-1})^s = A^{-s}$ $\quad (A^r)^s = A^{rs}$

Non-singular Matrix B has an inverse B^{-1}

(28a) *If* $A = BX$ *then* $X = B^{-1}A$

(28b) *If* $D = YC$ *then* $Y = DC^{-1}$

Problem 1104 Multiply M×G to calculate the elements of matrix C.

$C = M \times G$

$C = \begin{bmatrix} c_6 & c_5 & c_4 & c_3 & c_2 & c_1 & c_0 \end{bmatrix}$

$C = \begin{bmatrix} m_3 & m_2 & m_1 & m_0 \end{bmatrix} \times \begin{bmatrix} I_{4\times4} & | & R_{4\times3} \end{bmatrix}$

$C = \begin{bmatrix} m_3 & m_2 & m_1 & m_0 \end{bmatrix} \times \begin{bmatrix} 1 & 0 & 0 & 0 & 1 & 0 & 1 \\ 0 & 1 & 0 & 0 & 1 & 1 & 1 \\ 0 & 0 & 1 & 0 & 1 & 1 & 0 \\ 0 & 0 & 0 & 1 & 0 & 1 & 1 \end{bmatrix}$

Algebra

11.4 Rank of a Matrix

Definition of Rank A matrix has rank R when R is the largest integer for which *not all minors of order R are zero.*

Minor of order R The elements of a matrix are minors of order 1. They are determinants of order 1.

A minor of order R is a determinant of order R (page 61, Section 11.5).

Linear Dependence If r_p and r_q represent two rows of a matrix that has c columns, then $r_p + r_q$ is the c sums, element by element, of the two rows.

(29) *If* n_p, n_q, n_t *are three numbers so that* $n_p r_p + n_q r_q + n_t r_t = 0$,
then the three rows r_p r_q r_t *are linearly dependent.*

Rank The rank of a matrix is equal to the number of linearly *independent* rows in the matrix.

(30) *If matrix A has n rows and rank* $r < n$ *then there are* $n - r$
linearly dependent rows that can be expressed as sums of the
r linearly independent rows.

The rank of a matrix is difficult to determine. Determining the rank is difficult in the sense the calculations can be extensive, and difficult in the sense of finding a general format.

On the other hand, for example, a Vandermonde matrix has a special form *whose rank is known.* The terms of the rows are powers of the terms y_k in the first row.

$$(31) \quad V_{n \times n} = \begin{bmatrix} y_1 & y_2 & \cdots & y_n \\ y_1^2 & y_2^2 & \cdots & y_n^2 \\ \vdots & \vdots & & \vdots \\ y_1^n & y_2^n & \cdots & y_n^n \end{bmatrix} \quad rank = n, \; all \; y_i \; are \; distinct \; \& \; non \; zero$$

11.5 Determinants of a Matrix

The determinant of a matrix is

$$(32) \quad A = \begin{bmatrix} a_{11} & a_{12} & a_{13} \\ a_{21} & a_{22} & a_{23} \\ a_{31} & a_{32} & a_{33} \end{bmatrix} \rightarrow \Delta = \begin{vmatrix} a_{11} & a_{12} & a_{13} \\ a_{21} & a_{22} & a_{23} \\ a_{31} & a_{32} & a_{33} \end{vmatrix} \quad the\ determinant\ of\ A$$

Minor Δ_{pq} is formed by striking out row p and column q. The expansion by minors creates the n! terms of a n×n determinant. There are 6 terms (3!) in a 3×3 determinant.

$$(33) \quad \Delta = a_{11}\Delta_{11} - a_{12}\Delta_{12} + a_{13}\Delta_{13} \quad (expansion\ by\ row\ 1)$$
$$\Delta = a_{11}(a_{22}a_{33} - a_{23}a_{32}) - a_{12}(a_{21}a_{33} - a_{23}a_{31}) + a_{13}(a_{21}a_{32} - a_{22}a_{31})$$

The inverse of A is as follows so that $AA^{-1}=A^{-1}A=I$. Observe that row expansions are *columns* in A^{-1}.

$$(34) \quad A^{-1} = \begin{bmatrix} \dfrac{\Delta_{11}}{\Delta} & \dfrac{-\Delta_{21}}{\Delta} & \dfrac{\Delta_{31}}{\Delta} \\ \dfrac{-\Delta_{12}}{\Delta} & \dfrac{\Delta_{22}}{\Delta} & \dfrac{-\Delta_{32}}{\Delta} \\ \dfrac{\Delta_{13}}{\Delta} & \dfrac{-\Delta_{23}}{\Delta} & \dfrac{\Delta_{33}}{\Delta} \end{bmatrix}$$

The product AA^{-1} confirms the A^{-1} format.

$$(35) \quad AA^{-1} = \begin{bmatrix} a_{11} & a_{12} & a_{13} \\ a_{21} & a_{22} & a_{23} \\ a_{31} & a_{32} & a_{33} \end{bmatrix} \times \begin{bmatrix} \dfrac{\Delta_{11}}{\Delta} & \dfrac{-\Delta_{21}}{\Delta} & \dfrac{\Delta_{31}}{\Delta} \\ \dfrac{-\Delta_{12}}{\Delta} & \dfrac{\Delta_{22}}{\Delta} & \dfrac{-\Delta_{32}}{\Delta} \\ \dfrac{\Delta_{13}}{\Delta} & \dfrac{-\Delta_{23}}{\Delta} & \dfrac{\Delta_{33}}{\Delta} \end{bmatrix}$$

$$= \begin{bmatrix} a_{11}\dfrac{\Delta_{11}}{\Delta} - a_{12}\dfrac{\Delta_{12}}{\Delta} + a_{13}\dfrac{\Delta_{13}}{\Delta} & 0 & 0 \\ 0 & -a_{21}\dfrac{\Delta_{21}}{\Delta} + a_{22}\dfrac{\Delta_{22}}{\Delta} - a_{23}\dfrac{\Delta_{23}}{\Delta} & 0 \\ 0 & 0 & a_{31}\dfrac{\Delta_{31}}{\Delta} - a_{32}\dfrac{\Delta_{32}}{\Delta} + a_{33}\dfrac{\Delta_{33}}{\Delta} \end{bmatrix} = \begin{bmatrix} 1 & 0 & 0 \\ 0 & 1 & 0 \\ 0 & 0 & 1 \end{bmatrix} = I$$

12 Mathematical Induction

The mathematical induction method of *proof by induction* has many uses such as proving theorems, discovering new results, and providing relatively simple proofs of theorems obtained by other means.

The Principal of Mathematical induction A mathematical formula involving the positive integer n that is true for all positive integers provided that (1) the formula is true when n=1, and (2) the hypothesis that the formula is true for any n is sufficient to ensure that the formula is true for n+1.

Suppose a formula satisfies conditions (1) and (2). Then by (2) if the formula is true for n, it is also true for n+1.

Therefore if true for n=1 it is true for n=2,
      if true for n=2 it is true for n=3,
      and so for the for all values of n.

Example 1 Prove by induction that (1) $1^3 + 2^3 + 3^3 + ... + n^3 = \frac{1}{4}n^2(n+1)^2$

(2a) $n=1 \rightarrow 1^3 = 1$ *and* $\frac{1}{4}1^2(1+1)^2 = 1$

(2b) $n=2 \rightarrow 1^3 + 2^3 = 9$ *and* $\frac{1}{4}2^2(2+1)^2 = 9$

(2c) $n=n \rightarrow 1^3 + 2^3 + 3^3 + ... + n^3 = \frac{1}{4}n^2(n+1)^2$

(2d) $n=n+1 \rightarrow$ *add* $(n+1)^3$ *to both sides of* $=$

$1^3 + 2^3 + 3^3 + ... + n^3 + (n+1)^3 = \frac{1}{4}n^2(n+1)^2 + (n+1)^3$

$= \frac{1}{4}(n+1)^2[n^2 + 4(n+1)] = \frac{1}{4}(n+1)^2[n^2 + 4n + 4)] = \frac{1}{4}(n+1)^2[(n+2)^2]$ *qed*

Example 2 Prove by induction that (3) $1^2 + 2^2 + ... + n^2 = \frac{1}{6}n(n+1)(2n+1)$

(4a) $n=1 \rightarrow 1^2 = 1$ *and* $\frac{1}{6}1(1+1)(2+1) = 1$

(4b) $n=2 \rightarrow 1^2 + 2^2 = 5$ *and* $\frac{1}{6}2(2+1)(4+1) = 5$

(4c) $n=n+1 \rightarrow$ *add* $(n+1)^2$ *to both sides of* $=$

$1^2 + 2^2 + 3^2 + ... + n^2 + (n+1)^2 = \frac{1}{6}n(n+1)(2n+1) + (n+1)^2$

$= \frac{1}{6}n(n+1)(2n+1) + (n+1)^2 = \frac{1}{6}(n+1)[2n^2 + n + 6n + 6] = \frac{1}{6}(n+1)[(2n+3)(n+2)]$

Appendix

A1 Absolute Value

Definition The absolute value of x, denoted as |x|, is defined as follows.

$$|x| = \begin{cases} x & if \ x \geq 0 \\ -x & if \ x \leq 0 \end{cases}$$

However the absolute value itself is *always* positive. $|x| \geq 0$

The *absolute value* of x on the real number line is the distance from 0 to x.

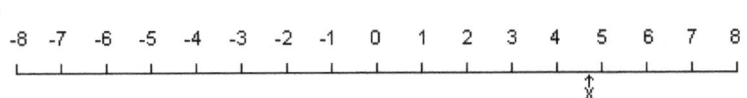

Since the absolute value of –q is q the solutions to the equation |x| = q are q and –q so that

If $|x| = 7$, then $x = 7$ or $x = -7$

If $|x| = q$, then $x = q$ or $x = -q$

and

If $|x| = |y|$, then $x = y$ or $x = -y$

Inequalities

If $q > 0$ and $|x| \leq q$, then $-q \leq x \leq q$

If $q > 0$ and $|x| \geq q$, then $x \leq -q$ and $x \geq q$

Examples

If $|3x-5| = |28|$, then $3x-5 = 28$ or $3x-5 = -28$ so that $x = \frac{33}{3} = 11$ or $x = -\frac{23}{3}$

If $|2x-5| < 6$, then $-6 < 2x-5 < 6$

Add 5 to each side $-6+5 < 2x-5+5 < 6+5 \rightarrow -1 < 2x < 11 \rightarrow -\frac{1}{2} < x < \frac{11}{2}$

If $|2x-5| = |x-4|$, then $2x-5 = x-4$ or $2x-5 = -(x-4)$

So that $x = 1$ or $x = 3$

Algebra

A2 Complex Numbers

The words complex and imaginary are potentially misleading, because complex numbers are not complicated and imaginary operators are not part of someone's imagination. Both words are labels: they are technical terms used to designate a class of numbers. A complex number z is represented by an ordered pair of real numbers x and y written as (x, y).

Multiplication by −1 and √−1 A number can be represented as a distance on a number line. We define steps to the right as positive so that distance AB=+4. Multiply +4 by −1 to get −4 that is the distance AC. Multiply AC by −1 to get back to AB. Clearly multiplication by −1 in effect *rotates* AB and AC by 180°.

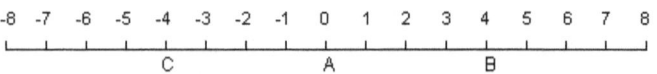

If +4 is multiplied by √−1 the result is 4√−1. Multiply 4√−1 by √−1 to get −4. Hence multiplication by √−1 two times rotates AB by 180°. And so multiplication by √−1 implements a 90° rotation of AB.

The world has agreed that numbers such as 4√−1 are *imaginary* numbers. To save writing √−1 is replaced by i in the mathematical literature.

Complex numbers The ordered pair (x_1, y_1) is a point in the (x, iy) plane that can be reached by starting from the origin, marching along the x-axis for a distance x_1, rotating $\pi/2$ radians, and marching parallel to the iy axis for distance y_1 (Figure A21a).

Working with ordered pairs (x, y) does not have much appeal, which is why the world adopted the well known alternative z=x+iy that is easier to work with.

In other words: taking our clue from the rotation operation we use i as a $\pi/2$ rotation operator. Then we say iy_1 is a vector we add to vector x_1 so that $z_1=x_1+iy_1$. This replaces the ordered pair (x_1, y_1). We say z is a complex number whose real part is x and whose imaginary part is y. Keep in mind that x and y are real numbers.

Figure A21 Complex numbers in Cartesian and polar coordinates

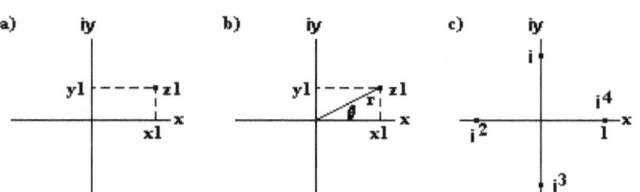

Polar coordinates: If r is the distance from the origin to the point z, then $x = r\cos\theta$, and $y = r\sin\theta$ (Figure A21b). See Euler relation below.

(1) $\quad z = x + iy = r\cos\theta + ir\sin\theta = re^{i\theta}$

(2) $\quad \tan\theta = \dfrac{y}{x}$ $\quad$ *so that* $\quad \theta = \tan^{-1}\dfrac{y}{x}$

Multiples of i Representing i as a $\pi/2$ rotation yields the same results as the $\sqrt{-1}$ representation (Figure A21c, Euler).

(3) $\quad i = e^{i\frac{\pi}{2}} = \cos\dfrac{\pi}{2} + i\sin\dfrac{\pi}{2} = 0 + i1 = i$

(4) $\quad i^2 = e^{i\frac{\pi}{2}2} = e^{i\pi} = \cos\pi + i\sin\pi = -1 + i0 = -1$

(5) $\quad i^3 = e^{i\frac{\pi}{2}3} = e^{i\frac{3\pi}{2}} = \cos\dfrac{3\pi}{2} + i\sin\dfrac{3\pi}{2} = -0 - i1 = -i$

(6) $\quad i^4 = e^{i\frac{\pi}{2}4} = e^{i2\pi} = \cos 2\pi + i\sin 2\pi = 1 + i0 = 1$

Addition The sum of complex numbers is found by adding the two x's, the two iy's, and factoring out i

$\quad z_1 + z_2 = (x_1 + iy_1) + (x_2 + iy_2)$

(7) $\quad z_1 + z_2 = (x_1 + x_2) + i(y_1 + y_2)$

Multiplication Find the product z_1z_2. To find it multiply z_1 and z_2, while *treating i as just another real number*. Then substitute -1 for i^2.

(8) $\quad z_1z_2 = (x_1 + iy_1)(x_2 + iy_2)$

$\quad\quad = x_1x_2 + x_1iy_2 + iy_1x_2 + iy_1iy_2$

$\quad\quad = x_1x_2 + iy_1iy_2 + iy_2x_1 + iy_1x_2$

$\quad\quad = x_1x_2 + i^2y_1y_2 + i(x_2y_1 + x_1y_2)$

$\quad\quad = (x_1x_2 - y_1y_2) + i(x_2y_1 + x_1y_2)$

113

Algebra

Subtraction Subtraction is defined as addition of positive and negative complex numbers.

(9) $\quad z_1 - z_2 = z_1 + [-z_2] = (x_1 + iy_1) + (-x_2 - iy_2)$
$$= (x_1 - x_2) + i(y_1 - y_2)$$

Division Division is facilitated by the complex conjugate concept, where i is replaced by −i.

$\quad$ If $z = x + iy$, then $\bar{z} = x - iy$

$\quad z\bar{z} = (x + iy)(x - iy) = x^2 - i^2 y^2 + ixy - iyx$

(10) $\quad z\bar{z} = x^2 + y^2 = r^2 = |z|^2 = |z| \times |z|$

$$\frac{z_1}{z_2} = \frac{z_1}{z_2} \times \frac{\bar{z}_2}{\bar{z}_2} = \frac{(x_1 + iy_1)(x_2 - iy_2)}{r_2^2} = \frac{x_1 x_2 - i^2 y_1 y_2 - ix_1 y_2 + iy_1 x_2}{r_2^2}$$

(11) $\quad \dfrac{z_1}{z_2} = \dfrac{x_1 x_2 + y_1 y_2}{r_2^2} + i\dfrac{x_2 y_1 - x_1 y_2}{r_2^2}$

Euler Relation (Figure A21b)

If $r = 1$ then $z = \cos\theta + i\sin\theta$

$\dfrac{dz}{d\theta} = -\sin\theta + i\cos\theta = i(\cos\theta + i\sin\theta) = iz$

(12) $\quad \dfrac{dz}{z} = id\theta$

Integrating $\quad \ln z = i\theta + constant$

If $\theta = 0$ then $z = 1$ so that $\ln 1 = i0 + constant$

However, $\ln 1 = 0$ so that $constant = 0$

$\therefore \ln z = i\theta \implies z = e^{i\theta}$

(13) $\quad e^{i\theta} = \cos\theta + i\sin\theta$

A3 Pascal's Triangle

The coefficients of a binomial expansion form Pascal's Triangle.

$(a+b)^1$				1	1		
$(a+b)^2$			1	2	1		
(14) $(a+b)^3$		1	3	3	1		
$(a+b)^4$	1	4	6	4	1		
$(a+b)^5$	1	5	10	10	5	1	

Answers to Most of the Problems

Problems 2 Fractions

An object is divided into 23 parts.
1. What is one part called? one twenty third
2. How is it written?   1/23

A number = 23/145.
3. Into how many parts is the original object divided? 145
4. How many parts of the original object does the fraction represent? 23

Find fractions equivalent to the following fractions that have a denominator of 128

5. 1/4	6. 3/8	7. 7/16	8. 5/32	9. 17/64
32/128	48/128	56/128	20/128	34/128

Find fractions equivalent to the following pairs of fractions that have a minimum common denominator

10. $\dfrac{1}{4}$ $\dfrac{3}{5}$ 11. $\dfrac{1}{14}$ $\dfrac{3}{21}$ 12. $\dfrac{7}{15}$ $\dfrac{5}{12}$ 13. $\dfrac{11}{16}$ $\dfrac{3}{15}$ 14. $\dfrac{1}{7}$ $\dfrac{3}{8}$

5/20 12/20 3/42 6/42   28/60   25/60   165/240 48/240 8/56 21/56

For each fraction find quotient plus remainder fraction.

15.	16.	17.	18.	19.	20.	21.	22.
8/8	9/8	10/8	11/8	12/8	13/8	14/8	15/8
1+0/8	1+1/8	1+2/8	1+3/8	1+4/8	1+5/8	1+6/8	1+7/8

23. 42/7	24. 22/7	25. 12/3	26. 14/3	27. 30/6	28. 33/6	29. 65/21
6+0/7	3+1/7	4+0/3	4+2/3	5+0/6	6+3/6	3+2/21

30. 27/11	31. 15/11	32. 19/4	33. 23/14	34. 45/9	35. 53/9	36. 27/5
2+5/11	1+4/11	4+3/4	1+9/14	5+0/9	5+8/9	5+2/5

Change to higher terms.

37. 2/7 to 28ths 38. 3/5 to 20ths 39. 9/8 to 32ths 40. 3/10 to 1000ths
8/28   12/20   36/32   300/1000
41. 11/6 to 72ths 42. 7/13 to 65ths 43. 2/9 to 27ths 44. 23/12 to 24ths
132/72   35/65   6/27   46/24

Algebra

Find the missing numbers.

45.	46.	47.	48.	49.	50.	51.
$\dfrac{42}{7}=\dfrac{\ }{28}$	$\dfrac{2}{3}=\dfrac{\ }{27}$	$\dfrac{9}{4}=\dfrac{36}{\ }$	$\dfrac{\ }{6}=\dfrac{25}{30}$	$\dfrac{30}{\ }=\dfrac{5}{3}$	$\dfrac{33}{7}=\dfrac{3}{\ }$	$\dfrac{8}{21}=\dfrac{48}{\ }$
168	18	16	5	18	77	126

Add the following pairs of fractions.

52.	53.	54.	55.	56.	57.
$\dfrac{8}{9}+\dfrac{1}{5}$	$\dfrac{3}{7}+\dfrac{1}{4}$	$\dfrac{5}{9}+\dfrac{3}{5}$	$\dfrac{9}{13}+\dfrac{2}{39}$	$\dfrac{2}{21}+\dfrac{3}{7}$	$\dfrac{4}{14}+\dfrac{1}{3}$
49/45	19/28	52/45	29/39	11/21	26/42

Convert to fractions and add.

58.	59.	60.	61.	62.	63.
$2\dfrac{7}{21}+4\dfrac{5}{14}$	$3\dfrac{1}{12}+5\dfrac{3}{18}$	$1\dfrac{3}{8}+3\dfrac{5}{6}$	$6\dfrac{7}{13}+1\dfrac{3}{26}$	$7\dfrac{7}{16}+1\dfrac{11}{12}$	$4\dfrac{3}{11}+1\dfrac{2}{9}$
281/42	297/36	125/24	199/26	113/48	544/99

Find the gcd of each pair of numbers.

64.	65.	66.	67.
255 *and* 153	336 *and* 280	136 *and* 255	105 *and* 168
51	56	17	21

Find the lcm of each pair of numbers.

68.	69.	70.	71.
255 *and* 153	336 *and* 280	136 *and* 255	105 *and* 168
765	1680	2040	840

Add the following pairs of fractions using the lcm

72.	73	74.	75
$\dfrac{13}{255}+\dfrac{7}{17}$	$\dfrac{4}{255}+\dfrac{5}{153}$	$\dfrac{9}{336}+\dfrac{2}{7}$	$\dfrac{13}{336}+\dfrac{3}{140}$
118/255	(12+25)/15×17×3	(9+96)/336	(65+36)/28×60

Find the lcm of

76.	77.	78.	79.
$\dfrac{1}{4}\ \dfrac{3}{5}\ \dfrac{7}{9}$	$\dfrac{2}{3}\ \dfrac{1}{15}\ \dfrac{9}{10}$	$\dfrac{5}{6}\ \dfrac{1}{2}\ \dfrac{1}{36}$	$\dfrac{3}{8}\ \dfrac{2}{9}\ \dfrac{1}{5}$
180	30	36	360

Subtract the fractions using lcm

80.	81.	82.	83.
$\dfrac{13}{85} - \dfrac{1}{17}$	$\dfrac{37}{51} - \dfrac{5}{255}$	$\dfrac{1}{14} - \dfrac{7}{168}$	$\dfrac{13}{56} - \dfrac{3}{140}$
8/85	180/255	5/168	59/280

84. Decrease the value of 3/4 by twelve thirty ninths. 69/156
85. Decrease the value of 7/10 by two fifths.   3/10
86. Decrease the value of 4/9 by one fifth.   11/45
87. Decrease the value of 3/7 by one tenth.   23/70
88. Decrease the value of 17/32 by three eighths. 5/32
89. Decrease the value of 5/16 by two sevenths.   3/112

Multiply the fractions. Reduce to lowest terms.

90.	91.	92.	93
$\dfrac{3}{22} \cdot \dfrac{55}{17}$	$\dfrac{18}{34} \cdot \dfrac{51}{28}$	$\dfrac{84}{13} \cdot \dfrac{39}{132}$	$\dfrac{28}{57} \cdot \dfrac{76}{7}$
15/34	27/28	21/11	304/57

94. Increase the value of 3/4 by four times.   3/1
95. Increase the value of 3/10 by two and one-half times. 3/4
96. Increase the value of 4/9 by five times.   20/9
97. Increase the value of 17/32 by two and two thirds times. 17/12
98. Increase the value of 4/9 by four and one half times.   2/1
99. Increase the value of 12/19 by one and one half times. 18/19
100. Decrease 1/3 to 1/7.   1/3 times 3/7
101. Decrease 3/5 to 1/10.   3/5 times 1/6
102. Decrease 17/32 to 1/16. 17/32 times 2/17

Divide the fractions. Reduce to lowest terms.

103	104	105	106
$\dfrac{\frac{3}{22}}{\frac{17}{55}}$	$\dfrac{\frac{18}{34}}{\frac{28}{51}}$	$\dfrac{\frac{13}{84}}{\frac{39}{132}}$	$\dfrac{\frac{57}{28}}{\frac{76}{7}}$
15/34	27/28	11/21	57/304

Problems 3 Decimals

1. The denominator of a Decimal fraction is a power of what number? 10
2. What is the purpose of the Decimal point?
   to mark separation of integer and fraction parts
3. A mixed number is the sum of two parts. What are their names?
   integer, fraction
4. The decimal point separates two parts of any number. What are their names?
   integer, decimal fraction

Convert to a decimal.

5.	6.	7.	8.	9.	10.
$\dfrac{3}{6}$	$\dfrac{3}{7}$	$\dfrac{5}{20}$	$\dfrac{13}{18}$	$\dfrac{17}{123}$	$\dfrac{5}{16}$
0.5	0.428571	0.25	0.7222	0.13821	0.3125

Convert to a decimal.

11.	12.	13.	14.	15.
$\dfrac{1}{6}$	$\dfrac{1}{11}$	$\dfrac{13}{45}$	$\dfrac{23}{54}$	$\dfrac{45}{111}$
0.1666	0.090909	0.2888	0.4259259	0.405405...

Convert the following phrases into decimals.
16. One-hundred-three one-thousandths 0.103
1. Thirty-nine one-thousandths   0.039
18. Nine one-hundredths   0.09

Expand the numbers into a sum of powers of ten. Omit the zero terms.

19.	20.	21.	22.
5.3	347.59	502.01	20001.0001

19. $5 \times 10^0 + 3 \times 10^{-1}$
20. $3 \times 10^2 + 4 \times 10^1 + 7 \times 10^0 + 5 \times 10^{-1} + 9 \times 10^{-2}$
21. $5 \times 10^2 + 2 \times 10^0 + 1 \times 10^{-2}$
22. $2 \times 10^4 + 1 \times 10^0 + 1 \times 10^{-4}$

23. What operation with what number moves a decimal point to the left?
   divide by ten

24. What operation with what number moves a decimal point to the right?   multiply by ten

25. Adding a leading zero to the integer part with value x changes the value to what?   same number x

26. Adding a trailing zero to the integer part with value x changes the value to what?   10x

27. Adding a leading zero to the fractional part with value x changes the value to what?   x/10

28. Adding a trailing zero to the fractional part with value x changes the value to what?   x no change

29. Write thirty seven hundredths in numerical format.   0.37
30. Write thirteen tenths in numerical format.   1.3
31. Write fifty seven thousandths in numerical format.   0.057
32. Write one hundred ninety nine ten thousandths in numerical format
        0.0199
33. Write one one millionth in numerical format.   0.000001
34. Find the repeating decimal produced by 1/6.   0.166666...
35. Find the repeating decimal produced by 1/11.   0.090909...
36. Find the repeating decimal produced by 2/13.   0.153846153846...
37. Find the repeating decimal produced by 77/123.   0.6260162601...
38. Find the repeating decimal produced by 2/11.   0.181818...
39. Find the repeating decimal produced by 12/123.   0.09756097560...

Use a calculator.
40. Find the non-repeating decimal produced by square root of 3.
        1.73205080...
41. Find the non-repeating decimal produced by p/2.   1.570796327...
42. Find the non-repeating decimal produced by log 2   0.301029996...
43. Find the non-repeating decimal produced by ln 2   0.693147181...
44. Find the non-repeating decimal produced by e^2 7.389056099...
45. Explain why the ratio of two integers cannot be irrational.

By definition they are rational. As a practical matter you can mark the point on the number line. E.g. 12/123. Divide distance from 0 to 1 into 123 parts, count 12 parts to right from 0.

Algebra

Find the sums and differences.

46	47.	48.	49.
$23.005 + 322.1$	$23.005 - 322.1$	$3007.1 + 1.3002$	$3007.1 - 1.3002$
345.105	-299.095	3008.4002	3005.7998

50.	51.	52.	53.
$100.01 + 1.01$	$100.01 - 1.01$	$5.06 + 3.067$	$5.06 - 3.067$
101.02	99.00	8.127	1.993

54.	55.
$90034.002 + 100.6$	$90034.002 - 100.6$
90134.602	89933.402

Find the products.

56.	57	58.	59.	60.
23.005×322.1	3007.1×1.3002	100.01×1.01	5.06×3.067	90034.002×100.6
7409.91050	3909.831420	101.01010	15.519020	9057420.601

61	62	63.	64.	65
$23.005 \div 322.1$	$3007.1 \div 1.3002$	$100.01 \div 1.01$	$5.06 \div 3.067$	$90034.002 \div 100.6$

61. $q = 0$, $r = 0.0714211919$
62. $q = 2312$, $r = 0.798031$
63. $q = 99$, $r = 0.019801980$
64. $q = 1$, $r = 0.649820672$
65. $q = 894$, $r = 0.9701988$

Problems 5

$$
\begin{array}{r}
x^2 - x\ -2 \\
x-2{\overline{\smash{\big)}\,x^3 - 3x^2 + 0x + 4}} \\
\underline{x^3 - 2x^2} \\
-x^2 + 0x + 4 \\
\underline{-x^2 + 2x} \\
-2x + 4 \\
\underline{-2x + 4} \\
0
\end{array}
$$

501

$$
\begin{array}{r}
3x^2 - 10x + 3 \\
x-1{\overline{\smash{\big)}\,3x^3 - 13x^2 + 13x - 3}} \\
\underline{3x^3 - 3x^2} \\
-10x^2 + 13x \\
\underline{-10x^2 + 10x} \\
3x - 3 \\
\underline{3x - 3} \\
0
\end{array}
$$

502

503

$$f(x) = 3x^3 - 22x^2y + 43xy^2 - 12y^3$$
$$f(y) = (3 - 22 + 43 - 12)y^3 = 12y^3 \neq 0$$
$$f(2y) = (24 - 88 + 86 - 12)y^3 = 10y^3 \neq 0$$
$$f(3y) = (81 - 198 + 129 - 12)y^3 = 0 \qquad (x - 3y) \text{ is a factor}$$

504

$$
\begin{array}{r}
3x^2 - 13yx + 4y^2 \\
x-3y\overline{\smash{)}\,3x^3 - 22yx^2 + 43y^2x - 12y^3} \\
\underline{3x^3 - 9yx^2} \\
-13yx^2 + 43y^2x \\
\underline{-13yx^2 + 39y^2x} \\
4y^2x - 12y^3 \\
\underline{4y^2x - 12y^3} \\
0
\end{array}
$$

$$
\begin{array}{r}
3x - y \\
x-4y\overline{\smash{)}\,3x^2 - 13yx + 4y^2} \\
\underline{3x^2 - 12yx} \\
-yx + 4y^2 \\
\underline{-yx + 4y^2} \\
0
\end{array}
$$

505

$$f(x) = x^7 - 5x + 3 \quad \rightarrow \quad f'(x) = 7x^6 - 5$$
$$f(0) = 3, f(1) = -1 \quad \rightarrow root\ between\ 0\ and\ 1$$
$$f(0.5) + h_1 f'(0.5) = 0.51 + h_1(-4.89) \quad \rightarrow \quad h_1 = 0.1043$$
$$f(0.6043) + h_2 f'(0.6043) = 0.0079 + h_2(-4.659) \quad \rightarrow \quad h_2 = 0.0017$$
$$f(0.6060) + h_3 f'(0.6060) = 0 - h_3(-4.65) \quad \rightarrow \quad h_3 = 0$$
$$root = 0.6060 \quad done$$

506

$$f(x) = x^5 - 3x^2 - 8 \quad \rightarrow \quad f'(x) = 5x^4 - 6x$$
$$f(0) = -8, f(1) = -10, f(2) = 12 \quad \rightarrow root\ between\ 1\ and\ 2$$
$$f(1.7) + h_1 f'(1.7) = -2.47 + h_1(-31.56) \quad \rightarrow \quad h_1 = 0.0783$$
$$f(1.7783) + h_2 f'(1.7783) = 0.297 + h_2(39.33) \quad \rightarrow \quad h_2 = -0.00755$$
$$root = 1.77075 \quad done$$

Algebra

Problems 6

601

$$f = 2x - y + 7 = 0 \qquad g = 3x + 4y - 6 = 0$$

$$4f + g = 11x + 22 = 0 \quad \rightarrow \quad x = -\frac{22}{11} = -2$$

$$y = 2x + 7 = -4 + 7 = 3$$

check $f = -2 \cdot 2 - 3 + 7 = 0 \qquad g = -3 \cdot 2 + 4 \cdot 3 - 6 = 0$

602

$$f = 2x - 3y - 10 = 0 \qquad g = 5x - 6y - 28 = 0$$

$$2f - g = -x + 8 = 0 \quad \rightarrow \quad x = 8$$

$$3y = 2x - 10 = 6 \quad \rightarrow \quad y = 2$$

check $f = 2 \cdot 8 - 3 \cdot 2 - 10 = 0 \qquad g = 5 \cdot 8 - 6 \cdot 2 - 28 = 0$

603

$$f = 6x - 10y - 8 = 0 \qquad g = -10x + 15y + 15 = 0$$

$$1.5f + g = -x + 3 = 0 \quad \rightarrow \quad x = 3$$

$$y = \frac{1}{15} \cdot 10x - \frac{1}{15} \cdot 15 = \frac{1}{15} \cdot 30 - \frac{1}{15} \cdot 15 = 2 - 1 = 1 \quad \rightarrow \quad y = 1$$

check $f = 6 \cdot 3 - 10 \cdot (1) - 8 = 0 \qquad g = -10 \cdot 3 + 15 \cdot 1 + 15 = 0 \quad qed$

604

$$3x^2 + 11x = 4 \quad \rightarrow \quad x^2 + \frac{11}{3}x = \frac{4}{3}$$

$$x^2 + \frac{11}{3}x + \left(\frac{11}{6}\right)^2 = \frac{4}{3} + \left(\frac{11}{6}\right)^2 = \frac{169}{36}$$

$$\left(x + \frac{11}{6}\right)^2 = \frac{169}{36} \quad \rightarrow \quad x = -\frac{11}{6} \pm \sqrt{\frac{169}{36}}$$

609

$$2x^2 + 2x - 1 = 0$$

$$x = -\frac{2}{4} \pm \frac{\sqrt{4+8}}{4} = -\frac{2}{4} \pm \frac{\sqrt{12}}{4} = -\frac{1}{2} \pm \frac{\sqrt{3}}{2}$$

610

$$3x^2 - 3x + 1 = 0$$

$$x = -\frac{3}{6} \pm \frac{\sqrt{9-12}}{6} = -\frac{1}{2} \pm \frac{\sqrt{-3}}{6} = -\frac{1}{2} \pm \frac{i\sqrt{3}}{6}$$

615

$$f = x + 2y - z = 6 \qquad g = 2x - y + 3z = -13 \qquad h = 3x - 2y + 3z = -16$$

$$f + h = 4x + 2z = -10$$

$$f + 2g = 5x + 5z = -20 \quad \rightarrow \quad k = x + z = -4$$

$$f + h - 2k = 2x = -2 \quad \rightarrow \quad x = -1, \qquad z = -4 - x = -3,$$

use f to get y -- $\quad 2y = 6 + z - x = 6 - 3 + 1 = 4 \quad \rightarrow \quad y = 2$

617

$$f : 3p - 2q + r = 6 \qquad g : 2p + 3q + 2r = -1 \qquad h : 5q - 4r = -3$$

from h : $4r = 5q + 3$

sub h into $4f$: $12p - 8q + 5q + 3 = 24 \quad \rightarrow \quad 12p - 3q = 21 \rightarrow 4p - q = 7$

sub h and $4f$ into $2g$: $4p + 6q + 4r = -2$

$$7 + q + 6q + 5q + 3 = -2 \quad \rightarrow \quad 12q = -12 \quad \rightarrow \quad q = -1$$

$$4p + 1 = 7 \quad \rightarrow \quad p = 6/4 = 3/2$$

$$4r = -5 + 3 = -2 \quad \rightarrow \quad r = -1/2$$

Problems 7 Simplify

1. $2^8 4^5 = 2^{18}$

2. $27^5 / 3^{11} = 3^4$

3. $25^{x+2} / 5^{x-1} = 5^{x+5}$

4. $9^{2m}(3^m)^{m+1} = 3^{m^2+5m}$

5. $\dfrac{4^2 2^{3n}}{8^{n+2}} = 2^{-2}$

6. $\dfrac{c^{x^2}}{c^{x^2(x+1)}} = \dfrac{1}{c^{x^3}}$

7. $\dfrac{(a^{2x-y})^{x+2y}}{(a^{2x+y})^{x-2y}} = a^{6xy}$

8. $\dfrac{x^{(a^2-9)}}{x^{a-3}} = x^{a^2-a-6}$

9. $\dfrac{a^{m-2n}a^{3(m+n)}}{a^{2m-n}} = a^{2m+2n}$

10. $\left(\dfrac{b^{2x-3}}{b^{2x+3}}\right)^{x+1} = b^{-6(x+1)}$

Find the values.

1. $81^{\frac{1}{2}} = 9$

2. $81^0 = 1$

3. $0^{\frac{1}{2}} = 0$

4. $64^{\frac{1}{4}} = 2^{\frac{3}{2}}$

5. $27^{\frac{1}{3}} = 3$

6. $27^{\frac{2}{3}} = 3^2$

7. $27^{\frac{4}{3}} = 3^4$

8. $16^{\frac{1}{4}} = 2$

9. $16^{\frac{3}{4}} = 2^3$

10. $\left(\frac{9}{25}\right)^{\frac{1}{2}} = \frac{3}{5}$

11. $\left(\frac{9}{25}\right)^{\frac{3}{2}} = \left(\frac{3}{5}\right)^3$

12. $0.04^{\frac{1}{2}} = \frac{1}{5}$

13. $0.216^{\frac{2}{3}} = \left(\frac{6}{10}\right)^2$

14. $(-8)^{\frac{1}{3}} = -2$

15. $\left(-\frac{1}{32}\right)^{\frac{2}{5}} = \frac{1}{2^2}$

16. $(-4)^3 = -64$

17. $7^{-2} = \frac{1}{49}$

18. $\left(\frac{2}{3}\right)^{-3} = \frac{2^{-3}}{3^{-3}} = \frac{3^3}{2^3}$

Algebra

Convert to positive exponents.

1. $x^{-2} = \frac{1}{x^2}$

2. $x^{\frac{3}{4}} x^{-\frac{1}{2}} = x^{\frac{1}{4}}$

3. $(x^{-\frac{2}{5}})^{-\frac{1}{4}} = x^{\frac{1}{10}}$

4. $(x^{-\frac{1}{2}})^{-\frac{5}{3}} = x^{\frac{5}{6}}$

5. $(-x^{-\frac{5}{6}})^{-\frac{1}{5}} = (-1)^{-\frac{1}{5}} x^{\frac{1}{6}} = -x^{\frac{1}{6}}$

6. $2x^{-1}y^{-2} = \frac{2}{xy^2}$

7. $\frac{3x^{-3}}{yz^{-4}} = \frac{3z^4}{x^3 y}$

8. $\frac{2x^{-1}y^4}{3^{-2}x^3 y^{-5}} = \frac{18 y^9}{x^4}$

9. $\frac{2^{-1}b^3 c^{-\frac{2}{3}}}{5b^{-\frac{1}{4}}c^2} = \frac{b^{\frac{13}{4}}}{10c^{\frac{8}{3}}}$

10. $\frac{3x^{-\frac{2}{5}}y^{-\frac{3}{2}}}{2^{-2}x^{-\frac{1}{2}}y^{-\frac{5}{6}}} = \frac{12x^{\frac{1}{10}}}{y^{\frac{4}{6}}}$

Convert denominator to 1.

1. $\frac{3x^2}{z^{-3}} = 3x^2 z^3$

2. $\frac{3a}{x^4 z^{-3}} = 3ax^{-4}z^3$

3. $\frac{x^2}{4y^{-\frac{2}{3}}} = x^2 4^{-1} y^{\frac{2}{3}}$

4. $\frac{x^{(a^2-9)}}{x^{a-3}} = x^{(a^2-a-6)}$

5. $\frac{c^{x^2}}{c^{x^2(x+1)}} = c^{-x^3}$

Simplify

4. $\frac{x^{-1}+y^{-1}}{y^{-2}-x^{-2}} = \frac{x^2 y^2}{x^2 y^2} \cdot \frac{x^{-1}+y^{-1}}{y^{-2}-x^{-2}} = \frac{xy}{1} \cdot \frac{y+x}{x^2-y^2} = \frac{xy}{x-y}$

9. $\frac{3+9x(9x^2+1)^{-\frac{1}{2}}}{3x+(9x^2+1)^{\frac{1}{2}}} = \frac{(9x^2+1)^{\frac{1}{2}}}{(9x^2+1)^{\frac{1}{2}}} \cdot \frac{3+9x(9x^2+1)^{-\frac{1}{2}}}{3x+(9x^2+1)^{\frac{1}{2}}}$

$= \frac{1}{(9x^2+1)^{\frac{1}{2}}} \cdot \frac{3(9x^2+1)^{\frac{1}{2}}+9x}{3x+(9x^2+1)^{\frac{1}{2}}} = \frac{3}{(9x^2+1)^{\frac{1}{2}}}$

Problems 8

801 $a^4 + 4a^3 b + 6a^2 b^2 + 4ab^3 + b^4$

802 $\frac{1}{8}b^3 - \frac{9}{4}b^2 x^3 + \frac{27}{2}bx^6 - 27x^9$

803

$e^{9x} + 9e^{7x} + 36e^{5x} + 84e^{3x} + 126e^x + 126e^{-x} + 84e^{-3x} + 36e^{-5x} + 9e^{-7x} + e^{-9x}$

804 Simplify

1. $\frac{5!}{3!} = 5 \cdot 4 = 20$

2. $\frac{9!}{6!} = 9 \cdot 8 \cdot 7$

3. $\frac{6! \cdot 8!}{7! \cdot 9!} = \frac{1}{7 \cdot 9}$

4. $\frac{4!+5!}{3! \cdot 4!} = \frac{1+5}{3!} = 1$

5. $\frac{5! \cdot 6!}{9!-7!} = \frac{5!}{(9 \cdot 8 - 1)7} = \frac{5!}{7! \cdot 7}$

6. $\frac{(n-1)!}{n!} = \frac{1}{n}$

7. $\dfrac{p!}{(p-2)!} = p(p-1)$

8. $\dfrac{2k!}{(2k)!} = \dfrac{2}{(2k-k+1)!}$

9. $\dfrac{n!}{(n-r)!}$

10. $\dfrac{(n-k-1)!}{(n-k+1)!} = \dfrac{1}{(n-k+1)(n-k)}$

11. $\dfrac{(n+1)!-n!}{n!+(n-1)!} = \dfrac{n!(n+1-1)}{(n-1)!n} = \dfrac{n^2}{(n-1)!}$

12. $\dfrac{[(2n+1)!]^2}{(2n)!(2n+2)!} = \dfrac{(2n+1)!}{(2n)!(2n+2)}$

13. *Show that $n!$, $n > 1$, is always an even number.*

14. *Show that* $\dfrac{n(n-1)(n-2)\cdots(n-r+1)}{r!} = \dfrac{n!}{r!(n-r)!}$

15. *Show that* $\dfrac{n!}{k!(n-k)!} + \dfrac{n!}{(k+1)!(n-k-1)!} = \dfrac{(n+1)!}{(k+1)!(n-k)!}$

805

1. $(x+y)^4 = x^4 + 4x^3y + 6x^2y^2 + 4xy^3 + y^4$

2. $(x-2y)^6 = x^6 - 12x^5y + 60x^4y^2 - 160x^3y^3 + 240x^2y^4 - 192xy^5 + 64y^6$

3. $(\frac{1}{2}x - 3y^3)^3 = \frac{1}{8}x^3 - \frac{9}{4}x^2y^3 + \frac{27}{2}xy^6 - 27y^9$

4. $(x^{\frac{1}{2}} + y^{\frac{1}{2}})^5 = x^{\frac{5}{2}} + 5x^2y^{\frac{1}{2}} + 10x^{\frac{3}{2}}y + 10xy^{\frac{3}{2}} + 5x^{\frac{1}{2}}y^2 + y^{\frac{5}{2}}$

806

1. $(x^{\frac{2}{3}} - \frac{1}{3}y^{-2})^{11} = x^{\frac{22}{3}} - \dfrac{11x^{\frac{20}{3}}}{3y^2} + \dfrac{55x^6}{9y^4} - \dfrac{55x^{\frac{16}{3}}}{9y^6} + \cdots$

2. $(x^{-3} + \frac{2}{3}x^{\frac{3}{2}})^{10} = \dfrac{1}{x^{30}} + \dfrac{20}{3x^{\frac{51}{2}}} + \dfrac{20}{x^{21}} + \dfrac{320}{9y^{\frac{33}{2}}} + \cdots$

3. $(x^4 - 2y^{-4})^{\frac{1}{4}} = x - \dfrac{1}{2x^3y^4} - \dfrac{3}{8x^7y^8} - \dfrac{7}{16x^{11}y^{12}} + \cdots$

4. $(8x^3 + 3y^2)^{\frac{2}{3}} = 4x^2 + \dfrac{y^2}{x} - \dfrac{y^4}{16x^4} + \dfrac{y^6}{96x^7} + \cdots$

807

1. $(\frac{1}{3}x^2 + y^{-2})^{10} = \cdots + \dfrac{40x^6}{9y^{14}} + \dfrac{5x^4}{y^{16}} + \dfrac{10x^2}{3y^{18}} + \dfrac{1}{y^{20}}$

2. $(\frac{1}{2}y^{-1} - ay^{\frac{1}{2}})^{12} = \cdots - \frac{55}{2}a^9y^{\frac{3}{2}} + \frac{33}{2}a^{10}y^3 - 6a^{11}y^{\frac{9}{2}} + a^{12}y^6$

3. $(\frac{2}{5}x^{\frac{2}{3}} + y^{\frac{1}{2}})^{11} = \cdots + \frac{264}{25}x^2y^{12} + \frac{44}{5}x^{\frac{4}{3}}y^{\frac{27}{2}} + \frac{22}{5}x^{\frac{2}{3}}y^{15} + y^{\frac{33}{2}}$

Algebra

808

1. 5th term of $(e^{2x} + e^{-2x})^{10}$ is $210e^{4x}$

2. 5th term of $(x^3 + 3y^3)^{\frac{4}{3}}$ is $\dfrac{5y^{12}}{3x^8}$

3. x^7 term of $(\frac{1}{2} + x)^{13}$ is $\frac{429}{16}x^7$

4. $x^{\frac{11}{2}}$ term of $(\frac{1}{4}x^{-1} + x)^{\frac{1}{2}}$ is $2x^{\frac{11}{2}}$

809

1. $(1.01)^9 = 1.0937$ 2. $(99)^4 = 96,064,000$ 3. $(103)^{\frac{1}{2}} = 10.149$

4. $(10)^{\frac{2}{3}} = 4.6416$ 5. $(1.03)^{-7} = 0.81309$

Problems 9

Convert exponential form $y = b^x$ to logarithmic form $x = \log_b y$.

1. $81 = 3^4$ 2. $2 = 8^{\frac{1}{3}}$ 3. $8 = 2^3$ 4. $10000 = 10^4$ 5. $\frac{1}{100} = 10^{-2}$

$\log_3 81 = 4$ $\log_8 2 = \frac{1}{3}$ $\log_2 8 = 3$ $\log_{10} 10000 = 4$ $\log_{10} \frac{1}{100} = -2$

6. $\frac{1}{64} = \left(\frac{1}{4}\right)^3$ 7. $y = 7^x$ 8. $y = b^3$ 9. $y = 10^x$ 10. $27 = b^x$

$\log_{\frac{1}{4}} \frac{1}{64} = 3$ $\log_7 y = x$ $\log_b y = 3$ $\log_{10} y = x$ $\log_b 27 = x$

Convert logarithmic form $x = \log_b y$ to exponential form $y = b^x$.

11. $5 = \log_2 32$ 12. $\frac{1}{4} = \log_{16} 2$ 13. $4 = \log_2 16$

$2^5 = 32$ $16^{\frac{1}{4}} = 2$ $2^4 = 16$

14. $6 = \log_{10} 1000000$ 15. $-3 = \log_2 \frac{1}{8}$

$10^6 = 1000000$ $2^{-3} = \frac{1}{8}$

16. $-3 = -\log_2 8$ 17. $z = \log_{10} y$ 18. $5 = \log_3 243$

$2^3 = 8$ $10^z = y$ $3^5 = 243$

19. $-3 = \log_{10} 0.001$ 20. $x = \log_3 81$

$10^{-3} = 0.001$ $3^x = 81$

Find $\log_{10}$ of the following numbers.

21. 100   22. 0.01   23. 1000   24. 1   25. 0.001

   2   −2   3   0   −3

26. 100000   27. 0.00001   28. 10   29. 0.1   30. 0.001

     5   −5   1   −1   −3

Solve for x.

31. $\log_{10} x = 5$ 32. $\log_x 16 = 4$ 33. $\log_2 x = 5$ 34. $\log_4 64 = x$

    $x = 10^5$ $x = 2$ $x = 2^5$ $x = 3$

35. $\log_{16} x = \frac{3}{2}$ 36. $\log_x 27 = \frac{3}{4}$ 37. $\log_{25} 625 = x$ 38. $\log_4 x = \frac{5}{2}$

    $x = 64$ $x = 81$ $x = 2$ $x = 32$

Find characteristic of $\log_{10} y$ for following y.

39. 7.234   40. 72.34   41. 0.7234   42. 72340   43. 7234×10^4

   0   1 -1 4   7

44. 0.007234   45. 72.34×10^{-6} 46. 723400   47. 0.72340×10^{-4}

   -3 -5 5 -4

If $\log_{10} y = 0.69897$, then y = 5. Use laws for log xy and log x/y to find log of the following numbers.

48. $\log 10y$ 49. $\log \frac{y}{10}$ 50. $\log 100y$ 51. $\log 1000y$ 52. $\log \frac{y}{1000}$

   1.69897 $-1 + 0.69897$ 2.69897   3.69897 $-3 + 0.69897$

53. $\log 2 (hint\ 2 = \frac{10}{5})$ 54. $\log \frac{2}{10}$ 55. $\log 200$ 56. $\log \frac{1}{2}$ 57. $\log \frac{100}{2}$

   $1 - 0.69897$ $0 - 0.69897$ 2.30103 -0.30103 1.69897

Log 10 = 1, Log 5 = 0.69897, 1−0.69897 = 0.30103. Find the antilog of the following numbers.

58. $1 - 0.69897$ 59. 2.69897   60. $-1 + 0.30103$ 61. $2 + 2.69897$

   $\frac{10}{5} = 2$ 500 $\frac{2}{10} = 0.2$ 100×500

62. $-1 + 0.69897$ 63. 4.30103   64. $2.30103 - 1.69897$ 65. $2.69897 - 1.30103$

   0.5   20000 $\frac{200}{50} = 4$ $\frac{500}{20} = 25$

Find the value of these expressions.

66. $\log 10^3$ 67. $\log(0.01)^4$ 68. $\log(0.001)^3$ 69. $\log 5^3$ 70. $\log 2^4$

   3   -8 -9 3×0.69897 4×0.30103

Expand as algebraic sum of terms.

71. $\log 3^2 7^3 5^7$ 72. $\log 9^{-1} 7^2$ 73. $\log 4^{\frac{1}{2}} 8^{\frac{1}{3}}$ 74. $\log 5^2 4^3$ 75. $3\log 5^2 7$

  $2\log 3 + 3\log 7 + 7\log 5$ $-\log 9 + 2\log 7$ $\frac{1}{2}\log 4 + \frac{1}{3}\log 8$ $2\log 5 + 3\log 4$ $6\log 5 + 3\log 7$

Problems 11

1104
The corresponding encoder is the G matrix. G implements the parity equations.

$C = M \times G$

$C = \begin{bmatrix} c_6 & c_5 & c_4 & c_3 & c_2 & c_1 & c_0 \end{bmatrix}$

$C = \begin{bmatrix} m_3 & m_2 & m_1 & m_0 \end{bmatrix} \times \begin{bmatrix} I_{4 \times 4} & | & R_{4 \times 3} \end{bmatrix}$

$C = \begin{bmatrix} m_3 & m_2 & m_1 & m_0 \end{bmatrix} \times \begin{bmatrix} 1 & 0 & 0 & 0 & 1 & 0 & 1 \\ 0 & 1 & 0 & 0 & 1 & 1 & 1 \\ 0 & 0 & 1 & 0 & 1 & 1 & 0 \\ 0 & 0 & 0 & 1 & 0 & 1 & 1 \end{bmatrix}$

Check:

$c_6 = m_3 + 0 + 0 + 0 = m_3$

$c_5 = 0 + m_2 + 0 + 0 = m_2$

$c_4 = 0 + 0 + m_1 + 0 = m_1$

$c_3 = 0 + 0 + 0 + m_0 = m_0$

$c_2 = m_3 + m_2 + m_1 + 0 = r_2$

$c_1 = 0 + m_2 + m_1 + m_0 = r_1$

$c_0 = m_3 + m_2 + 0 + m_0 = r_0$

INDEX

www.ingramcontent.com/pod-product-compliance
Lightning Source LLC
Chambersburg PA
CBHW051710170526
45167CB00002B/614